Sustainable Energy Development

Sustainable Energy Development: A Multi-Criteria Decision Making Approach discusses sustainable energy development, the main path for achieving carbon neutrality, and the use of multi-criteria decision making (MCDM) in assessing energy transition in both operational and socio-political forms. It proposes ways to support responsible decision making toward sustainable energy in key areas such as power distribution, household energy, and transportation. The authors have developed frameworks and tools to help choose sustainable energy options like renewable energy technologies, energy efficiency improvements, policies, and how to promote them in different communities. The book includes several case studies focused on electricity, district heating, transport sectors in the European Union (EU), and decision making in the household sector.

Features

- Analyzes the connection between sustainable energy development and the transition toward a carbon neutral society.
- Compares and discusses advanced MCDM tools to support sustainable energy options.
- Develops new frameworks of indicators for the assessment of sustainable regional and national energy system planning, and provides practical illustrative examples in various energy sectors.
- Provides policy implications when promoting sustainable energy development.
- Presents case studies on the applications of multi-criteria tools to support sustainable energy options in different energy sectors.

Readers interested in gaining insight into leading trends in energy efficiency and sustainability, such as academics, researchers, graduate students, and professionals interested in sustainable energy and energy producers, city planners, policy makers, and more, will benefit from the topics and frameworks discussed in this book.

Sustainable Energy Development

A Multi-Criteria Decision Making Approach

Indre Siksnelyte-Butkiene and Dalia Streimikiene

CRC Press
Taylor & Francis Group
Boca Raton London New York

CRC Press is an imprint of the
Taylor & Francis Group, an **informa** business

Designed cover image: © iStock Photo

First edition published 2024
by CRC Press
6000 Broken Sound Parkway NW, Suite 300, Boca Raton, FL 33487-2742

and by CRC Press
4 Park Square, Milton Park, Abingdon, Oxon, OX14 4RN

CRC Press is an imprint of Taylor & Francis Group, LLC

Library of Congress Cataloging-in-Publication Data
Names: Siksnelyte-Butkiene, Indre, author. | Streimikiene, Dalia, author.
Title: Sustainable energy development : a multi-criteria decision making approach /
Indre Siksnelyte-Butkiene, Dalia Streimikiene.
Description: First edition. | Boca Raton : CRC Press, [2023] |
Includes bibliographical references and index. |
Summary: "This book discusses sustainable energy development, the main path for achieving carbon neutrality, and the use of multi-criteria decision making (MCDM) in assessing energy transition in both operational and socio-political forms. It proposes ways to support responsible decision making towards sustainable energy in key areas such as power distribution, household energy, and transportation. Intended for both academics and professionals involved in policymaking, the authors have developed frameworks to help choose sustainable energy options like renewable energy technologies, energy efficiency improvements, policies, and how to promote them to different communities"– Provided by publisher.
Identifiers: LCCN 2022060976 (print) | LCCN 2022060977 (ebook) |
ISBN 9781032346496 (hardback) | ISBN 9781032355108 (paperback) |
ISBN 9781003327196 (ebook)
Subjects: LCSH: Renewable energy sources. |
Energy transition–Decision making. | Multiple criteria decision making.
Classification: LCC TJ808 .S55 2023 (print) | LCC TJ808 (ebook) |
DDC 621.042068/4–dc23/eng/20230123
LC record available at https://lccn.loc.gov/2022060976
LC ebook record available at https://lccn.loc.gov/2022060977

ISBN: 978-1-032-34649-6 (hbk)
ISBN: 978-1-032-35510-8 (pbk)
ISBN: 978-1-003-32719-6 (ebk)

DOI: 10.1201/9781003327196

Typeset in Times
by Newgen Publishing UK

Contents

Introduction ix

Chapter 1   Sustainable Energy Development Policy Implications 1

    1.1   Sustainable Development and Energy 1
    1.2   Sustainable Energy Development Concept and Its Free
          Dimensions 4
    1.3   Historical Development of Sustainable Energy Paradigm 9
    1.4   The Main Topics of Sustainable Energy Development
          Research 14
    1.5   Policies and Measures to Encourage Sustainable
          Energy Development 19

Chapter 2   Multi-Criteria Decision Aiding and the Governance of
               Sustainability 25

    2.1   Decision Makers and Stakeholders in the Energy Sector 25
    2.2   Participation and Deliberation as a Tool for Inclusive
          Policy Making 27
    2.3   Multi-Criteria Decision Making as a Tool to Deal with
          Compromises 29
    2.4   Selection of Indicators 32
    2.5   Comparative Evaluation of the Most Popular MCDM
          Techniques 37

Chapter 3   Multi-Criteria Decision Making for Regional and
               National Planning 45

    3.1   Sustainable Regional and National Energy
          System Planning 45
    3.2   Practical Examples of Multi-Criteria Analysis Application 47
        3.2.1   Framework to Monitor Policy Progress by
                Achieving Energy Policy Goals 47
                3.2.1.1   Energy Policy Context 48
                3.2.1.2   Indicators for Comparative
                          Assessment of EU Energy Policy
                          Priorities Implementation 51
                3.2.1.3   MCDM Technique 56
                3.2.1.4   Multi-Criteria Assessment Results 59

3.2.2   Framework to Monitor Electricity Sector
         Sustainability 66
         3.2.2.1   Energy Policy Context 66
         3.2.2.2   Indicators for Comparative Assessment
                    of Electricity Sector Sustainability............67
         3.2.2.3   MCDM Tool..70
         3.2.2.4   Multi-Criteria Assessment Results.............71
3.2.3   Framework to Monitor Heating Sector
         Sustainability 74
         3.2.3.1   Energy Policy Context and
                    Overview of EU Heating Sector.................74
         3.2.3.2   Indicators for Comparative Assessment
                    of Heating Sector Sustainability.................78
         3.2.3.3   MCDM Implementation..............................83
         3.2.3.4   Multi-Criteria Assessment Results.............84
3.2.4   Framework to Monitor Transport Sector
         Sustainability 88
         3.2.4.1   Energy Policy Context 89
         3.2.4.2   Indicators for Comparative Assessment
                    of Road Transport Sustainability................90
         3.2.4.3   MCDM Technique.......................................94
         3.2.4.4   Multi-Criteria Assessment Results.............95
Appendix 3.1 Abbreviations of the EU Member States 100

Chapter 4   Multi-Criteria Decision Making for Sustainable Transport
            Development..111

4.1   Sustainable Transport Development 111
4.2   Selection of Instruments for Sustainable Transport
       Decision Making...112
       4.2.1   Public Transport Planning Issues..............................117
       4.2.2   Sustainability Assessment..119
       4.2.3   Logistics Issues...120
       4.2.4   Transport Project Selection Issues.............................122
       4.2.5   Transport Policy Issues...123
4.3   Criteria for Sustainable Transport Decision Making 124

Chapter 5   Multi-Criteria Decision Making for Sustainable Energy
            Development in Households..133

5.1   Sustainable Energy Development in Households 133
5.2   Selection of Instruments for the Renewable Energy
       Technologies Selection 135
       5.2.1   The Main Types of Renewable Energy
                Technologies in Households 135
       5.2.2   The Application of MCDM Instruments....................137

5.3   Criteria for the Assessment of Renewable Energy
        Technologies 140
5.4   Selection of Instruments for the Building Insulation
        Materials Selection.. 145
5.5   Criteria for the Assessment of Building Insulation
        Materials 148

Conclusions.. 165

Index... 169

Introduction

The issues of sustainable energy development are the core topics today in many countries around the world. The importance of such problems as energy dependency, energy security, or renewable energy development became clear, not only in political documents, but also in the everyday lives of energy users. Promotion of sustainable energy development is the main way to achieve carbon neutrality by the mid-century. This book proposes the main ways to support responsible decision making toward energy sustainability in such important energy sectors as electricity, district heat, and transport. The book develops frameworks and tools to select sustainable energy options like renewable energy technologies, energy efficiency measures and policies and measures to promote them.

The main issues of sustainable energy are linked to climate change, as energy consumption is the main source of human-created greenhouse gas (GHG) emissions, which are responsible for 75.6 percent of worldwide GHG emissions. The energy sector includes energy use in transportation, households, electricity, heat production, manufacturing, construction, agriculture, and services. Though there are significant achievements in GHG emission reductions from the energy sector due to energy efficiency improvements and the use of renewable energy sources, nevertheless, GHG emissions are growing in transportation and households, which are the major energy consumers and GHG emission sources of this sector. As there is a huge GHG emission reduction potential in these sectors, the book focuses on applications of MCDM for dealing with sustainable energy development issues in these sectors.

The monograph aims to address the current policy priorities regarding the low carbon energy transition and creation of climate neutral society, and to reveal the suitability of different MCDM methods to deal with contradictory issues of sustainable energy development. The book discusses theoretical issues of sustainable energy development, including policy implications to promote sustainable energy development; it provides methodological recommendations for indicator selection, it overviews the main sectors toward low-carbon energy transition and presents frameworks for multi-criteria analysis; and it provides illustrative examples of multi-criteria analysis, taking into account policy objectives for sustainable development.

The first chapter is intended to analyze and systematize the literature on sustainable energy development issues. A detailed analysis of the scientific literature allows for systematizing each group of sustainable energy development challenges and policies necessary to implement established targets. Based on a detailed analysis of sustainable energy development issues, the main areas are identified as: promotion of renewable energy and improvement of energy efficiency, both having positive impacts on GHG emission reduction.

The second chapter indicates the main actors in the energy sector, discusses participation and deliberation as tools for inclusive policy making, presents the main requirements for the selection of indicators in order to perform assessments in a sustainable way, and introduces methodological recommendations for energy study assessments based on the Bellagio Sustainability Assessment and Measurement Principles (STAMP). Also, comparative evaluation of the most popular and valid MCDM techniques for dealing with sustainable energy issues is provided in this chapter.

The third chapter justifies the importance of sustainable regional and national energy system planning and provides practical illustrative examples of multi-criteria analysis applications to monitor energy policy progress and assess electricity, heating, and transport sector sustainability by developing three case studies. The created MCDM frameworks and assessment procedures were applied for the measurement of achievements made in the European Union (EU) member states in electricity, district heating, and transport sectors.

The fourth chapter provides an in-depth overview of MCDM techniques applied for sustainable decision making in the transport sector throughout the world. The overview of methods is categorized according to the application areas that include: sustainability assessment, transport policy, public transport planning, project selection, and logistics categories. Finally, the thematic areas for criteria selection regarding the issues of sustainable transport decision making have been developed and systematized.

The fifth chapter is an overview of the scientific literature that has used MCDM methods as a key tool for decision making in households. The analysis consists of two problematic questions: renewable energy technology selection and insulation materials selection. The main criteria of sustainability assessment in the household sector were systematized for renewable energy micro generation technologies and housing renovation options ranking.

Conclusions and policy implications based on advancement in MCDM tools and their applications for decision making in favor of sustainable energy development are provided at the end of the book.

1 Sustainable Energy Development Policy Implications

1.1 SUSTAINABLE DEVELOPMENT AND ENERGY

Economic development is a somewhat more wide approach than only economic growth. And with the Afterword of the Gro Harlem Brundtland report, "Our Common Future" in 1987, environmentally sustainable economic development became the main paradigm of economic expansion (World Commission on Environment and Development, 1987). Therefore, the paradigm of sustainable development was first of all positioned to the coherence between people and nature as a promise for future world development.

When comparing a sustainable development paradigm with other approaches of economic development, it is obvious that a sustainable development approach considers the quality of the environment as more significant than other issues linked to economic development. Sustainable development requires the emergence of a new economy, which recognizes the limits of the biosphere and assures the equilibrium between ecological systems and their social implications for humans.

Sustainable development thinkers try to find answers to the same questions as other development economists, but sustainable thinkers include their concern for future generations. According to sustainable development theorists, economic growth must be treated as a complex idea, one which also encourages the responsibility of humans for all that is happening, and it reflects the necessary changes in space and time to secure a sustainable economy – which is the goal of long-term development of the planet. Therefore, the switch from a consumer economy to an economy oriented to the harmony between people and the environment is necessary in order to achieve sustainable development.

The concept of sustainable development has become very popular in recent years. The famous British environmentalist David Pearce called the concept "sustainable development" and "sustainability" as well as their derivatives, such as "sustainable agriculture", "sustainable energy", "sustainable economic development" as the most modern words of the twentieth century.

The importance of energy was already recognized when the idea of sustainable development was acknowledged, first in the United Nations (UN) issuing "Our Common Future" report in 1987 (World Commission on Environment and Development, 1987).

DOI: 10.1201/9781003327196-1

However, energy was initially mainly mentioned in the context of emission reduction and increasing energy security.

The concept of sustainable energy development was first introduced in 2000, and the environmental, economic, and social impacts of energy systems were documented (United Nations Development Programme et al., 2000). As a new paradigm, sustainable energy development was used to define the role of energy in realizing sustainable development targets (United Nations Development Programme et al., 2000). The UN's "Sustainable Development Goal (SDG) 7", linked to affordable and clean energy for all has further strengthened the inevitability of energy for achieving sustainable development (United Nations, 2022). Over time, sustainable energy development has transformed to a wide ranging international policy goal connected with some of the major developmental challenges facing the world. Gradually, the importance of this concept has increased, with fossil fuels depleting and rising concerns due to climate change (United Nations Development Programme et al., 2000).

Sustainable energy can be defined as energy produced and utilized in modes that help to achieve long term human development, in all its social, economic, and environmental dimensions or production and usage of energy in modes that promote long-term human well-being and balance with ecosystems (Gunnarsdottir et al., 2021).

The energy system can be treated as unsustainable because of very important equity, environmental, economic, and geopolitical fears having importance for future world development. The main issues of unsustainability of energy supply and use are the following (Spittler et al., 2019):

- Electricity and modern-day fuels are not accessible to many people around the word and energy inequity and injustice has important moral, political, and practical implications due to globalization and interconnectedness of world;
- The energy system is not dependably reliable to support widespread economic growth due to energy supply security issues. The lack of access to modern commercial energy sources and insecurity and unreliable energy supplies has direct negative impact on economic development and low productivity levels of many countries;
- Energy production and use are associated with negative environmental impacts on local, regional, and global levels and have negative effects on human health and well-being for future generation as well.

Therefore, most important for addressing sustainable energy development issues is discovery of ways and modes to increase energy supply affordability and security for humans by addressing the negative environmental impacts of energy use. This achievement would allow also for fostering economic growth and social progress (Diesendorf, Elliston, 2018).

Poverty reduction is indeed a key SDG, as the World Bank now praises. However, sustainable development cannot be achieved now in all world countries as environmental and social costs of production have started to rise faster than the benefits of production as a result of GDP growth. Such a type of economic growth makes some people poorer (Jenkins et al, 2017). It is also because even real economic growth cannot increase prosperity on the margins by producing goods and services that meet mostly relative, rather than absolute, human needs. If the welfare of people depends largely on relative income, overall growth automatically nullifies its effect on well-being. The evident solution is to restrict such non-economic growth in rich countries to allow at least temporary continued growth in poor countries; however, this solution is rejected by the ideology of globalization, which can only promote global growth (Sovacool et al, 2016). Then, it is necessary to support national and international policies that aim to tax adequately for resources use in order to limit the size of the economy relative to the ecosystem, and to generate revenue for public purposes. These policies must be based on economic theory, which includes key environmental concepts, including sustainable development. This effective national policy needs to be protected from cost pass-throughs and reduction of competition, and must promote globalization (Hussain et al., 2017).

Energy is at the core of almost all SDGs – from intensifying access to modern energy services and switching to non-polluting cooking fuels, from diminishing subsidies for fossil fuels to cutting fatal air pollution that annually causes the premature deaths of millions of people around the planet. One of SDG's – SDG7 shows the importance of energy for sustainable development and seeks to guarantee access to affordable, reliable, sustainable, and modern energy sources for all people (United Nations, 2022).

Hence, energy is a vital component of economic and social development. Together, traditional forms of energy supply, generation, distribution, and usage are associated with environmental damage, negative human health impacts, including reduction of quality of life and threatening the Earth's ecosystems. In addition, the affordability of energy services is extremely uneven. There are over 7.88 billion people, most of whom live in rural areas in developing countries and do not have access to a commercial energy supply. Although lack of access to a modern energy supply is among the most important obstructions to sustainable development, better access to commercial energy services has direct positive effects on poverty reduction. Access to affordable and adequate modern energy services should increase considerably to advance the living standards of the growing population around the world, especially in the least developed countries (Hashim, Ho, 2011).

The current liberalization and transformations in energy sector are mainly driven by increasing economic globalization, will drive societies to more cost-effective energy markets. These transformations provide an opportunity to guarantee that public benefits linked to energy services required for sustainable development are duly taken into account in evolving energy market reforms. Indeed, one of the main goals of market liberalization in many states in the 1990s was to decrease governmental subsidies and to attract private capital for necessary energy investments. Other goals were to promote innovations, cost-effectiveness, and management efficiency in the energy sector due to market liberalization (Dincer, 2000).

Energy markets provide for greater economic efficiency, lower prices for customers, improved risk mitigation, and the improvement of competitiveness and economic growth expansion. However, markets are not perfect and market failures are evident in energy sector as well. The size and time horizons of energy installations and relationship of energy supply with public goods and external costs like environmental pollution have historically led to creation of policies and measures necessary to deal with market failures. Competitive markets can allocate resources in more efficient ways than controlled systems, but the market does not take into account external costs of energy supply and use. For this reason policies and measures are necessary to correct market distortions and integrate external costs or external benefits of public goods related to energy production and consumption (Bataille et al., 2016).

The challenge facing both developed and developing countries calls for strong political will and pledge to modernization and the use of energy-efficient, environmentally friendly, cost-competitive energy technologies (Diezmartinez, 2021). Environmentally friendly technological opportunities are known, and they can guarantee a sustainable energy future possible for all countries (Demirbas et al., 2009). Safeguarding satisfactory access to energy for all humans in an environmentally, socially, and economically sustainable mode will require significant struggles, major investments, improvement of available institutions and policy systems (Curtin et al., 2019).

The UN is strongly supporting the goal of sustainable development, which means satisfying the needs of the present population without conceding the possibilities of future generations to satisfy their own needs. The role of energy as a means to this end has been recognized at every main UN conference, starting with the UN Conference on Environment and Development, or the Rio Summit, since 1992 (United Nations, 1992). However, the current energy systems still do not meet the basic needs of all humans around the world and, if continuing with ordinary business practices, this situation could also jeopardize the prospects of all future generations (Curtin et al., 2019).

Sustainable energy is energy supply and consumption in a way that supports sustainable human development in the long term, taking all social, economic and environmental issues into consideration. Sustainable energy refers supply and consumption of energy resources in ways that promote, or are at least compatible with, long-term human well-being and ecological equilibrium among economic development and nature. In the following section, the evolution of the sustainable energy concept during the last decades is analyzed.

1.2 SUSTAINABLE ENERGY DEVELOPMENT CONCEPT AND ITS FREE DIMENSIONS

As the definition of sustainable development recognizes the economic, social, and environmental impacts, sustainable energy development also needs to take into account the economic, social, and environmental needs of the current and future populations. Sustainable energy should be based on clean, renewable sources of energy, which can renew by themselves and cannot be depleted. There is no one interpretation or understanding of this concept. The most popular definitions of sustainable energy include environmental, economic, and social dimensions (Figure 1.1).

The World Energy Council (2019) has proposed the "Energy Trilemma" approach and developed the "Energy Trilemma Index" to measures three dimensions of energy sustainability: Energy Security, Energy Equity and Environmental Sustainability (Figure 1.2).

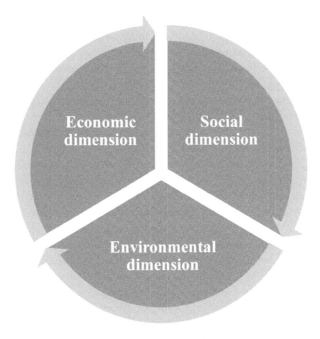

FIGURE 1.1 Sustainable energy development dimensions.

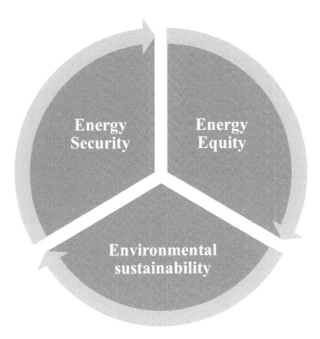

FIGURE 1.2 Energy Trilemma concept.

As one can see from Figure 1.1 and Figure 1.2 the same economic (energy security), social (energy equity) and environmental (environmental sustainability) dimensions are represented in the Energy Trilemma concept.

Factually, the concept of sustainable energy development in the beginning was focused on pollution and energy security issues. Later, the sustainable energy development concept was extended to cover more social and economic problems (Brew-Hammond, 2012).

Therefore, it is obvious that sustainable energy development is a multifaceted and multidimensional concept. It was first introduced in the UN's World Energy Assessment, and the relationship between sustainable development and energy production and use were highlighted (United Nations Development Programme et al., 2000).

So, there are three main dimensions of sustainable energy: economic, social, and environmental. The economic dimension of sustainable energy is mainly linked with energy supply security and the competitiveness of energy sector. A least cost energy planning approach should be applied for the developing of energy scenarios. The economic dimension of sustainability also deals with economic development, efficient use of energy, and security of energy supply. The social issues of energy sustainability traditionally comprise access to affordable and reliable energy for all people. However, not just the cost-effectiveness of the energy sector and energy supply security issues are important from the point of view of the economic dimension. Though energy is primarily a driver of wealth and people's well-being as reliable and affordable energy supply services is the major production factor in all sectors of economy, the environmental dimension of sustainable energy is of key importance (Cherp, Jewell, 2014). Energy production should not exceed the carrying capacity of the ecosystems (United Nations Development Programme et al., 2000). Therefore, in the center of this concept, as in the case of the sustainable development concept, is environmental sustainability (Spading-Fecher et al., 2005).

The most important environmental dimension of energy sustainability concentrates on greenhouse gas (GHG) emission reduction, atmospheric pollution, and the negative impact of this pollution on ecosystems includes negative health impacts of toxic waste, water consumption and pollution, depletion of fossil fuel resources (Jefferson, 2000). Energy resources with lower ecological impacts are often termed as green or clean sources of energy. The only renewable energy that can be considered really sustainable are the green and clean energy sources (Banos et al., 2011). The most common renewable energy sources (RES) are: wind, solar, hydro, bioenergy, and geothermal energy (Ullah et al., 2021). Bioenergy is energy created from biological material like wood, straw, waste water, manure, biogas and other agricultural residual or by products (Lund, 2007). Nevertheless, some RES might have negative environmental impacts, such as cutting down forests to produce biofuels or raising livestock on agriculture land that could be used for growing food (Evans et al., 2009).

Though nuclear power can also be considered as a low-carbon energy carrier, there is big concern about it due to severe accidents like Chernobyl, nuclear proliferation risks, and nuclear waste disposal problems. Therefore, the role of all non-renewable energy sources is quite controversial in sustainable energy development concept.

Switching from dirty fossil fuels like lignite or coal to cleaner fossil fuels like natural gas is positive trend as it allows to reduce GHG emissions (because of the lower carbon content of natural gas). However, it can cause delay in further low carbon energy transitions to RES. There are carbon capture and storage technologies which allows the removal of CO_2 emissions from coal, fuel oil, and other fossil fuels power plants; however, these technologies have a lot of their own challenges linked to CO_2 transportation and storage options (Zhang et al., 2009).

Some countries have a limited understanding of sustainable development and meeting a low or zero-carbon energy future (Martinot et al., 2002). Currently, the "green deals" are being proposed for a green recovery and a more sustainable future, such as the "European Green Deal" and the "Green New Deal" in the United States (Shivakumat et al., 2019; Olabi, Abdelkareem, 2022).

However, while a better understanding of sustainability is still lacking, it is difficult to push towards a more sustainable future without the risk of repeating past mistakes. Therefore, there is a need for a clearer and more structured understanding of sustainability and sustainable energy development concepts (Wang et al., 2019).

One way of identifying the relevant sustainability issues in the energy sector is a bottom-up approach, where stakeholders are engaged to decide what is sustainable development and what it means for them. The relevant stakeholders and communities can be actively involved in determining what a desirable energy future should be (Hosein Ghorashi et al., 2021). This approach is very useful, as challenges of sustainable energy development can vary significantly around the world based on geographic, climate, and other conditions (Sen, Ganguly, 2017).

Therefore, context-specific analysis of sustainable energy development issues supports improved and informed decision making and policy development (Cieplinski et al., 2021). In addition, increased public participation can provide for increased general support for public actions and policies (Hu et al., 2018).

Developing ways to monitor progress and inform actions towards sustainable development are very important for international institutions as well as for countries (United Nations Department of Economic and Social Affairs, 2007). Sustainability indicators are applied for monitoring progress towards sustainable development (United Nations, 1992). The International Atomic Energy Agency et al. (2005) has developed the most comprehensive framework for sustainable energy assessment consisting in economic, social, and environmental indicators. The interrelations between these indicators are shown in Figure 1.3.

Multiple different indicators for sustainable energy development vary in both purpose and quality (Davidsdottir et al., 2021). These efforts have been stymied by ambiguities in the concept of sustainable energy development, especially in the local context (Najam, Cleveland, 2003). Several limitations have been identified in existing energy indicators for sustainable development, such as: inconsistent results, a singular perspective, "backward-looking", and overemphasizing economic aspects (Davidsdottir et al., 2021). Furthermore, the various disagreements among scientists on methodological approaches and whether, for instance, stakeholders should be involved during sustainable energy indicator selection for monitoring progress towards sustainable energy development, have proven challenging (Fuso Nerini et al., 2018).

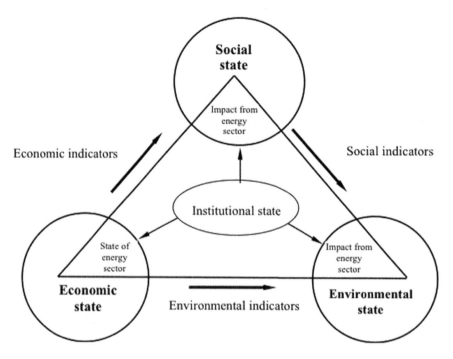

FIGURE 1.3   Interrelations between sustainable energy dimensions.

Climate change is one of the major challenges of the 21st century (United Nations, 2015). Climate change mitigation requires significant changes in energy systems due to the strong linkages between GHG emissions and combustion of fossil fuels. Therefore, for climate change mitigation purposes, the global energy systems should be de-carbonized. Energy systems must move away from the fossil-fuel dependent systems towards systems based on RES, like biomass, solar, wind and so forth, in order to meet the Paris Agreement.

There are three main interlinked issues in addressing sustainable energy development: increased use of renewables, energy efficiency improvements, and GHG emission reductions. These three interlinked issues represent environmental dimensions of sustainable energy development. Social dimension of sustainable energy development is represented by energy affordability and access to energy. The economic dimension of sustainable energy development is mainly linked to competitiveness of energy systems and security of energy supply (Naruala, Reddy, 2015).

The economic, social and environmental issues of sustainable development are also closely interlinked. Increasing the use of renewables and energy efficiency improvements provide a lot of opportunities, not only for GHG emission reduction but also for increase of energy supply security. This is because renewables and energy efficiency improvements reduce energy import dependency by increasing opportunities for distributed energy generation and energy access in remote areas. At the same time, with policy support measures, energy affordability can be improved in renewable energy systems due to economically viable energy prosumption and energy cooperatives options. Therefore, the measures to support the use of renewables and improvements in energy efficiency play a major role (Mao et al., 2015).

1.3 HISTORICAL DEVELOPMENT OF SUSTAINABLE ENERGY PARADIGM

The most commonly used sustainable development paradigm was first introduced in 1987 in the Brundtland report (World Commission on Environment and Development, 1987). This report stressed the role of energy for implementing sustainable development aim, though the sustainable energy concept was not fully elaborated during 1980s.

Agenda 21 was adopted at the UN Conference on Environment and Development in 1992, and sustainable development became a major policy goal on all levels, including global, national and local. Agenda 21 mentioned sustainable energy development in relation with sustainable development, mainly in Chapter 9, which is dealing with atmospheric protection (United Nations, 1992).

The link between the atmospheric pollution, GHG emissions, and energy are supplementary emphasized in 1992 in the context of UN Framework Convention on Climate Change (UNFCCC) and the Kyoto Protocol, signed in 1997. The UNFCCC provided a new attitude to sustainable energy development by emphasizing the importance of the energy sector on reducing GHG emissions.

Therefore, the extended sustainable energy development paradigm emerged in that time after adoption of Agenda 21 and the UNFCCC Kyoto protocol. Energy was considered a necessity for sustainable development mainly because of addressing climate change related issues, energy security, and depleting fossil fuel resources. Also, during this time, such topics like fast penetration of RES, increase of energy efficiency, and addressing negative environmental impact of energy supply and the role of energy in social progress were additionally deliberated. Nevertheless, a comprehensive approach to sustainable energy development had not yet been suggested. In addition, the concept of sustainable use and management of renewable energy resources was very new during these times.

The UN General Assembly in 1997 took the first steps towards development of a sustainable energy development agenda and transformed it finally into SDGs in 2015 (United Nations General Assembly, 2015). Although energy development was a constant topic on the world policy agenda during 1990s, the real comprehensive concept of sustainable energy development was only introduced in 2000. A sustainable energy development paradigm was introduced as a new development paradigm in the UN World Energy Assessment report issued in 2000 (United Nation Development Programme, 2000). One of the most important changes in the sustainable energy development paradigm was the more holistic approach applied to sustainable energy's development definition by taking into account all possible economic, environmental, or social impacts of energy supply, production, and consumption. The focus has been on a more environmentally sound and more diverse

mix of energy sources, equity in of energy access, improvement in energy efficiency, and on stressing the importance of viability of energy systems necessary for
meeting current and future demand. In addition, UN Millennium Declaration the
Millennium Development Goals (MDGs) were introduced in 2000 (United Nations
Department of Public Information, 2000). Though, the declaration did not provide
energy targets or emphasis of energy in achieving the MDGs and did not identify
sustainable energy development as a policy theme, some MDGs were closely linked
with energy consumption and the addressing of environmental problems of sustainable development, including GHG emission reduction targets. Also, it is necessary to stress that energy issues were seen in a more isolated way and were being
treated as separate from other SDGs. So, exclusion of direct sustainable energy
targets from the MDGs required a more clear elicitation of energy role in fostering
social and economic progress in the world. In 2002 the Johannesburg Declaration on
Sustainable Development was adopted, further pursuing SDGs for countries worldwide without clear delineation of sustainable energy development prospects (United
Nations, 2002).

Energy was the first and major subject in discussion about sustainable development at the ninth session of the UN Commission on Sustainable Development held in
2011 (UN Commission on Sustainable Development, 2011). This meeting prepared
major discussion topics for the second World Summit on Sustainable Development
to be held in 2012 and stressed the importance of international cooperation to shape
international actions for a more sustainable energy future (United Nations, 2022).
During this meeting the significance of energy access in reducing energy poverty and
achieving all MDGs was stressed (Modi et al., 2005). The urgent need to switch from
current unsustainable energy production and consumption patterns was highlighted.
Accordingly, the main attention was placed on the two main objectives of sustainable
energy development: to improve access to energy and to encourage cleaner energy
system development. Nevertheless, the agreement on the objectives or actions of
sustainable energy development was not reached during this meeting due to lack of
experience in international cooperation on energy issues. Consequently, the main outcome of this meeting in terms of sustainable energy development was introducing it
on the higher level in a sustainable development agenda and the highlighting of the
need for an institutional body, given the cross-cutting nature of the issue.

During the following years the establishment of various institutions related
to sustainable energy development were established in UN and various energy
initiatives introduced by these established UN bodies dealing with sustainable
energy issues. UN-Energy was created in 2004 as an inter-agency body to help
countries in transition towards sustainable energy and coordination of various
energy related actions in the different UN bodies. The Energy and Climate Change
initiative was commended by the UN Industrial Development Organization, and the
UN Development Programme initiated assessments on Energy and Environment
actions in UN Member States. In addition, the UN Division for Sustainable
Development started to deal with sustainable energy issues as well. In 2009, the
International Renewable Energy Agency was established. This is an intergov-
ernmental body aiming to support use of RES. The creation of these bodies and

various initiated activities by them show that sustainable energy development became a topical issue on the international policy agenda. Moreover, their formation acknowledged the increasing importance of energy in achieving various SDGs and the urgency of introducing sustainable consumption and production principles in the energy sector.

In 2010, after ten years of established Millennium Development, a special resolution was issued whereby energy was presented as necessary in achieving the goals. Energy access, efficiency, and sustainability were emphasized in this resolution in particular. In 2011 the Sustainable Energy for All (SE4ALL) initiative was endorsed by the UN Secretary General. The significance of addressing sustainable energy development problems for achieving the MDGs was stressed in developing the SE4ALL initiative. The goal of this initiative was to deliver by 2030 sustainable energy for all, by stressing the importance of such issues as energy access and energy security, energy use efficiency, and renewable energy resources.

Agenda 2030, linked to SDG, was ratified in 2015. The objective of the SE4ALL initiative was transferred to SDG7: "Ensure access to affordable, reliable, sustainable and modem energy for all". Therefore, just with the introduction of a SDG linked to the energy sector was real acknowledgement of a sustainable energy development paradigm. Therefore, the concept of sustainable energy development was clearly placed in the center of sustainable development paradigm as SDG7.

In 2006, the International Energy Agency published the first report on Energy Technology Perspectives, which is focused on the importance of energy technologies in addressing SDGs. The report stressed the importance of a cleaner and sustainable energy future with affordable energy prices for achieving world sustainable development priorities In 2017, sustainable development scenarios were linked to necessary actions to meet the SDGs by 2030. Therefore, activities of International Energy Agency and other institutional bodies like the UN Department of Economic and Social Affairs, European Environment Agency, World Energy Council, International Atomic Energy Agency, and so forth, had significant impact on further evolution of sustainable energy development concept.

So, the most important period for evolution of sustainable energy development concept was the 1980s and 1990s decades. Before these decades a very narrow view was dominating energy development by emphasizing emissions reduction and energy security issues. Moreover, energy issues were seen as isolated from other sustainable development issues. During recent decades, energy development was analyzed from a broader perspective by addressing the potential social and economic effects of energy sector development. Recently, the importance and central role of energy in enhancing social and economic development was acknowledged worldwide (Huang et al., 2008).

It is necessary to stress that the necessary changes in unsustainable energy production and consumption patterns are emphasized now in the evidence of climate change problems in recent debates on sustainable energy development. Currently, energy accessibility, affordability, low carbon energy transition, energy efficiency and creation of carbon neutral economy and society are the main topics in energy policy agendas everywhere. In this way a sustainable energy development paradigm has obtained a more comprehensive definition with increased stress on its importance for delivering the major SDGs including the fight against poverty, hunger, inequality, climate change, and so forth.

So, the sustainable energy paradigm has been changing through the years due to important events linked to the sustainable development concept analyzed above. The impact of progress of a sustainable development concept and related events on evolving of sustainable energy development concept are summarized in Table 1.1.

TABLE 1.1
The Impact of Sustainable Development Concept on Energy Sector

Year	Major Events Having Effect on Sustainable development paradigm	Sustainable Development Concept	Implications for Sustainable Energy development
1972	UN Conference on the Human Environment, 5–16 June 1972, Stockholm; Stockholm Declaration and Action Plan for the Human. Environment.	Limits of fossil fuel usage and their implications on environment.	Environmental impact of energy systems; energy security issues due to depletion of fossil fuels.
1987	The Brundtland Commission officially dissolved in 1987 after releasing Our Common Future, also known as the Brundtland Report.	Sustainable development concept.	Conserved energy costs and energy supply curves were addressed.
1988	World Meteorological Organization and the United Nations Environment Programme (UNEP) created Intergovernmental Panel on Climate Change for the provision of scientific information for governments that they can use to develop climate policies.	Climate change issues as the major issues of sustainable development.	Energy is one of the most important drivers of climate change due to fossil fuel combustion.
1992	UN Conference on Environment and Development in Rio and Agenda 21st. The UNFCCC was established.	Energy and climate research were merged.	Special report on GHG Emission Scenarios and Global Energy Perspectives were published with contribution of energy scholars.

TABLE 1.1 (Continued)
The Impact of Sustainable Development Concept on Energy Sector

Year	Major Events Having Effect on Sustainable development paradigm	Sustainable Development Concept	Implications for Sustainable Energy development
2000	UN Development Programme World Energy Assessment.	Energy sector and its challenges for sustainability.	The concept of sustainable energy was adopted.
2001	9th Session report of UN Commission of Sustainable Development.	Energy for sustainable development became a central topic at UN.	Among 58 indicators of sustainable development 3 sustainable energy development indicators were included.
2002	World Summit on Sustainable Development.	Relationship between energy and socio-economic development were addressed. Cross-scale energy systems impacts (national, regional and global).	Energy is central for sustainable development.
2010	UN Millennium Development Goals.	Energy is central for achieving MDG by 2010 and implementing sustainable development.	Energy is central for achieving all MDGs.
2011	UN SE4ALL initiative and Global Action Agenda.	UN initiative to promote sustainable energy development.	Initiative to positively transform the world's energy systems containing 11 Action Areas. It has established framework for identifying the opportunities necessary for prompt innovation and changes of energy sector.

(*continued*)

TABLE 1.1 (Continued)
The Impact of Sustainable Development Concept on Energy Sector

Year	Major Events Having Effect on Sustainable development paradigm	Sustainable Development Concept	Implications for Sustainable Energy development
2012	The UN Conference on Sustainable Development – or Rio+20.	The main focus on political document which contains clear and practical measures for implementing sustainable development.	The Conference also took forward-looking decisions on a number of thematic areas, including energy, food security, cities, etc.
2015	Agenda 2030 and SDGs, adopted by the UN as a universal call to action to end poverty, protect the planet, and ensure that by 2030 all people enjoy peace and prosperity.	Energy was set as a integral part of sustainable development with SDG7 (affordable and clean energy for all).	There are many energy interlinked SDG like affordable and clean energy (SDG7), industry, innovation and infrastructure (SDG9), sustainable cities and communities (SDG11), responsible and sustainable consumption (SDG12), climate action (SDG13).

Source: Created by authors.

It is clear from the summary provided in Table 1.1 that sustainable energy development has progressed to become a fundamental international policy goal, and that energy has become an essential factor to achieve sustainable development. The thematic analysis of the main areas linked to sustainable energy development was performed based on systematic literature review in the next section of this chapter.

1.4 THE MAIN TOPICS OF SUSTAINABLE ENERGY DEVELOPMENT RESEARCH

Energy is central to achieving sustainable development as energy is the main driver of economic progress and allows all sectors of economy to function well and be innovative. Energy is central for social progress and allows for achieving prosperity, well-being, quality of life, and also addresses equity issues. Energy is the major source of GHG emissions and other classical and atmospheric pollutants that have significant impact on air, land, and water pollution that bring about negative human health impacts. All these issues are interlinked, as negative effects of environmental impacts are transferred to social and economic spheres. The positive impacts of economic growth due to energy are also transferred into progress in the social sphere. Although economic, social, and environmental issues of sustainable energy development are closely interlinked, several thematic areas can be distinguished.

The main topics of sustainable energy development based on systematic literature review are summarized in Table 1.2.

TABLE 1.2
The Main Topics of Sustainable Energy Development

Topic	Description
The effect of energy system on climate change and policies to mitigate these impacts.	The effect of the energy system on climate change is a very broad and popular topic in sustainable energy development debate. The effect is assessed through life cycle approach including resource extraction, transportation, plant building, production, consumption and disposal.
The negative environmental impacts of energy systems and policies to mitigate these impacts.	The negative effect of energy systems like the classical atmospheric pollutants not directly effecting climate change but providing local negative impact on environment (atmosphere, water, land, human health). The life cycle and life cycle external costs approach is also been followed.
The climate change effect on energy systems and implications for resiliency.	The impact of climate change on the renewable energy resources development linked to natural or geographical conditions (solar radiation, wind speed, precipitation level etc.) and adaptation of energy systems to climate change are analyzed.
The limited fossil fuel resources and related policy propositions.	The limits of fossil fuels and necessary transformations of energy systems due to availability, security of supply and high cost.
The limitations and intermittency of renewable energy resources and policy propositions.	The temporal availability of RES influencing the future renewable energy because of low availability and high costs.
The effect of energy system on socio-economic development.	The human health impacts of energy options including mortality, morbidity, affordability and poverty and policies to support positive effect and mitigate negative effect are developed.
Trade-offs and synergies between specific energy sector development goals.	Energy has important social, environmental and economic effects. Specific goals are targeting specific sustainability dimension, therefore it is necessary to trade between specific social, environmental and economic goals of energy sector development by defining their complimentary agreement or disagreement.

(continued)

TABLE 1.2 (Continued)
The Main Topics of Sustainable Energy Development

Topic	Description
The effect of energy systems development on global development and vice versa.	The energy system development of a country or region can influence development of other countries like distribution of scarce resources, shortages in energy supply, climate effects, etc.
Energy security in energy supply and policies to increase it.	Energy security deals with the security of energy supply and production, and such insecurities like availability and affordability linked to high energy prices. The security of energy supply addresses the short-term and long-term supply. Energy availability is linked to the short-term intermittency of energy supply. Policies and measures to increase energy security and reduce energy import dependency are developed.
Sustainable energy consumption issues and policies to promote it.	The use of RES and energy efficiency measures as well as behavioral changes towards energy saving can provide a lot of benefits for society and allow for reducing negative environmental impacts of fuel combustion. Also these policies allow to increase energy affordability and reduce energy poverty.
Energy affordability and policies to increase it.	Energy affordability measures the affordability of energy services. Energy affordability shows the deprivation in a specific domain of consumption. It sometimes overlaps with energy poverty measurements and policies to mitigate energy poverty by increasing energy affordability.
Energy poverty and policies to mitigate it.	Energy poverty refers to energy vulnerability add shows the lack of distributional equity linked to limited ability of individuals to use energy. It can be linked to income poverty or poor housing conditions including low insulation of buildings, health problems related to cold homes or very high temperatures in buildings during summer etc. all having negative impacts on the life quality.
Energy equity and policies to enhance it.	Energy equity shows an ability of country to ensure the entire access to affordable, fairly priced and abundant energy supply in residential and other end use sectors. Policies and measures to enhance energy equity are analyzed and developed.
Energy justice and policies to enhance it.	Energy justice concept is related with application of justice principles to energy policy, energy production, distribution and consumption, energy activism, energy security and climate change. Energy justice can be considered as distributional, recognition and procedural. The main issues are evaluation where injustices emerge and how to deal with energy injustice. Policies and measures to address energy injustice are developed.

Source: Created by authors.

All the topics described in Table 1.2 can be summarized in the following main broad areas of sustainable energy development research: international energy economy issues; sustainable energy supply; energy security; sustainable energy consumption and energy affordability.

Five themes of sustainable energy development are linked to overarching goal of sustainable energy development, that is, energy systems need to develop in order to reach a sustainable energy development path.

There are some overlapping and close linkages between sustainable energy development themes in scientific literature. For example, Energy Indicators for Sustainable Development developed by IAEA and other international institutions reflect linkages between economic, social, and environmental themes linked to sustainable energy development and provide cross-cutting issues of sustainable energy development.

It is important to stress that sustainable energy development goals can be very different for various countries due to diverse priorities linked to distinct energy systems and discrete economic development levels. The developing countries are facing quite distinct energy sector improvement problems compared to developed nations. The availability of fossil and renewable energy resources is a major factor driving energy sector development in terms of energy production and consumption.

However, for all countries there is the same broader goal of energy sector development to provide secure and affordable energy supply necessary for economic growth and social progress by minimizing negative environmental impacts of energy production and consumption – as energy is necessary for achieving all other social, environmental, and economic goals of sustainable development. Hence, notwithstanding the different situations and energy-related problems of specific countries, the same overall goal of sustainable energy development is common for all countries to promote sustainable development taking into account economic, social, and environmental dimensions.

In line with these main themes of development (international energy economy issues; sustainable energy supply, energy security, sustainable energy consumption, and energy affordability), sustainable energy should improve world-wide with intergenerational equity by supplying modern energy services to all. Therefore, these modern energy services should first of all be accessible and affordable to all people by contributing to their well-being and an increase in living standards. Of course, environmental impact of energy services is supposed to be very limited.

To achieve this goal, it is necessary to transform the current energy system by switching to low carbon or RES that are environmentally friendly and can solve energy resource scarcity problems, including high energy import dependency and lacking of energy resources. The new energy technologies require to switch from fossil and nuclear fuels to RES. However, this transition is only possible if renewable energy technologies are economically viable. Another very important issue linked to modern energy technologies is increase in energy efficiency. Nevertheless, increase in energy generation efficiency stipulated by use of modern innovative and efficient technologies also requires taking into account energy use efficiency issues. As with increased use of energy efficient technologies, the rebound effect can be expected (Sorrel et al., 2009). Therefore, sustainable energy consumption should be promoted by increasing energy efficiency awareness and stimulating behavioral changes towards sustainable consumption behavior patterns. Sustainable energy consumption patterns allow for achieving enormous energy savings and related massive GHG emissions avoidance. They also allow for saving financial resources by reducing energy bills and providing for win-win solutions for consumers, suppliers, and general society as well. There are other important energy efficiency improvement methods in households, like renovation of residential buildings, installation of micro generation renewable energy technologies, use of energy efficient appliances, and an increase in importance and popularity of zero carbon buildings in residential, commercial, and the public sector.

Renewable energy sources have intermittency problems, and additional infrastructure (including huge energy storage capacities) is necessary to ensure secure and reliable supply of energy (Zhu, Xu, 2020).

Low carbon energy transition requires taking into account energy affordability and reduction of energy poverty, as RES are more expensive or have higher upcoming costs in comparison with investments necessary for fossil fuels, therefore energy equity and energy justice issues should be treated with high concern in prompting zero carbon transformations of energy systems (Kim, 2021).

It is evident from scientific literature, that problems of access to modern energy services characteristic to developing countries will be less pressing issues in the future. More important problems of sustainable energy development seen in the future are linked to development of zero carbon energy supply systems, especially in the transport sector. In household sector energy prosumption, energy storage and development of new business models will be the major concerns of sustainable energy research (Olabi, Abdelkareem, 2022).

The sustainable utilization of energy resources in terms of energy security and competition for the land among food and other services, like tourism and recreation, will require more attention from scientific community policy makers. There are many other barriers hampering fast penetration of renewables like the "not in my back yard" (NIMBY) phenomena, and financial, institutional, and behavioral barriers (Owen, 2006; Hu et al., 2018).

Future research in the sphere of sustainable energy needs to address the main policies and measures necessary to promote sustainable energy development (Wichairi, Sopadang, 2018). In policy analysis, the sustainability assessment of policies and measures can play very important roles as policies to promote penetration of RES, increase of energy efficiency, and reduce energy poverty require being assessed on their impact, on economic, social, and environmental issues of sustainable energy development, that is, increased security of energy supply, energy affordability and justice, and reducing GHG emissions and other negative environmental impacts (Nerini et al., 2018).

1.5 POLICIES AND MEASURES TO ENCOURAGE SUSTAINABLE ENERGY DEVELOPMENT

Since the current path of energy development is still not fully compatible with the key targets of sustainable development, governments and international institutions therefore need to concentrate on appropriate policies and measures to initiate necessary changes to ensure a sustainable energy development path (Rosenthal et al., 2018).

The main policies and measures in support of sustainable energy development should center on ensuring wider energy access, promotion of energy efficiency in energy generation and consumption, acceleration of renewable energy penetration in the energy markets, and intensifying the use of renewables in the households sector (International Energy Agency, 2017).

These policies and measures usually aim at market failures hampering achievement of sustainable energy development goals, like penetration of renewables and GHG emission reduction. Therefore, measures promoting innovations, overcoming market barriers, and imperfections for protecting significant public benefits are necessary in the energy sector.

It is possible to define the main sustainable energy policy objectives, these are:

- To safeguard the continued access to modern, high quality, secure, clean, and affordable energy services for all people in the current and future world;
- To guarantee security and reliability of energy supply in the short-, medium-, and long-term periods for all sectors;
- To ensure a decrease in GHG emissions and all environmental pollution and negative health impacts linked to energy production, supply, distribution, and consumption;
- To develop well-established modern energy grids and networks tailored to increased operating and cost efficiency of energy systems;
- To support continued energy efficiency improvement in energy supply, production and consumption, especially in economies in transition;
- To promote use of innovative, clean, energy efficient and economically viable energy technologies.

These main policy issues can be summarized in the following categories:

- Energy market opening, liberalization, and promotion of economic efficiency of energy sector;
- Energy pricing, subsidization and internalization of externalities;
- Energy availability and affordability;
- Security of energy supply and energy import dependency;
- Energy efficiency and intensity;
- Renewable energy technologies.

In order to achieve a sustainable energy future, it necessary to achieve the following objectives:

- Build stronger partnerships between governments, regional authorities, municipalities, energy producers, energy consumers, non-governmental organizations,
and financial institutions to facilitate a common understanding of the issues,
challenges, and constraints related to sustainable energy development;
- Increase the quantity, quality, and dissemination of statistics, information, and
analyses on links between energy on the one hand, and quality of life, economic development, and the environment on the other, so as to foster a holistic
approach and to improve energy decision making at all levels of society;
- Enhance institutional and individual human resource capacities related to
energy matters in order to move the concept of sustainable energy development
beyond its conceptual formulation to the implementation stage by overcoming
policy incompatibilities and analytical fragmentation.
- Develop norms, standards, conventions, guidelines and best practices as well as
voluntary agreements relating to energy production, use, and trade in order to
facilitate the economic integration of all countries in the region, and their energy
systems, and to substantially reduce energy-related environmental impacts.
- Establish an enabling environment, such as appropriate legal, regulatory,
and policy frameworks, for promoting investments (especially private sector
investments) for securing energy supplies; integrating energy networks and
systems; enhancing energy efficiency; accelerating the development and implementation of renewable energy resources; and promoting research and development on existing technologies and developing new environmentally sound and
economically viable energy production and use technologies.
- Make greater use of economic instruments, such as emission trading and tax-
ation, in addition to the continued use of regulatory measures applied in a
domestic setting with national boundaries to alter consumer behavior and meet
environmental objectives.
- Persist and in some cases accelerate, the implementation of further economic
reforms and restructuring of the energy sectors of transition countries, concomi-
tant with provision of assistance through institutional capacity development,
training, and market formation activities by Western countries and institutions.

The most important and urgent sustainable energy policies in developing countries
need to address energy reliability, affordability, and health and environmental impacts
of energy use.

For developed countries the most important sustainable energy development policies and measures are linked to overcoming renewable energy and energy efficiency
improvement barriers. Such measures like voluntary or mandatory standards applied
for buildings, energy appliances, vehicles, and other labelling schemes can be very
useful, as they allow for better informing customers and to shape their preferences
towards sustainable energy consumption options. Also, green public procurement
policies are effective for achieving economy of scale in production of efficient
appliances, vehicles, and so forth. Training in new renewable and energy efficiency
technologies is also very useful to increasing market penetration of these technologies. Financial support measures and credit mechanisms are necessary to help consumers meet high investment costs of renewable and energy-efficient technologies.

Governments need to establish performance characteristics of clean technologies by providing priorities for sustainable energy technologies in the energy market. The main regulatory measures for supporting renewable energy are mandates requiring a certain percentage of energy consumption to be supplied by renewable sources, requiring priority of energy supply to grids for independent renewable energy producers.

Legal and regulatory measures are necessary to encourage private companies to invest in renewable energy technologies, and allow for mitigation of major risks linked to such types of investments. Political stability and the rule of law, including necessary institutional environments are essential to facilitate investments in new clean energy technologies.

Direct support by the government can help to support technological innovation in the energy sector by reducing costs and providing tax incentives, thereby supporting research and development in sustainable energy technologies and other energy market transformation initiatives.

The human, natural, and technological resources in developing countries need to be further advanced to ensure low carbon energy sector transition, and these countries need assistance from developed countries in terms of technology transfer, financial resource provision, and human capacity building. The private sector could gain access to new markets by such transfers of new technologies to developing countries. For technology sharing and capacity building in developing countries, the training provided by public agencies, research institutes, and international bodies is essential.

It is necessary to stress that the current globalization process allows to transfer finances, technologies, and energy flow from one country to another. International harmonization of environmental taxes and emissions trading, implementation of international energy efficiency standards for products, machinery and vehicles are also very important steps towards a sustainable energy future. The concerted climate change mitigation actions in the energy sector are also obvious due to UNFCCC and Paris Agreement pledges.

Therefore, the main challenges of sustainable energy development need to be addressed by countries worldwide in partnerships by establishing cooperation in sustainable energy development, enabling international organizations, multilateral financial institutions, and civil society to take an active part in achieving secure, available, affordable, and clean energy for all.

REFERENCES

Banos, R., Manzano-Agugliaro, F., Montoya, F.G., Gil, C., Alcayde, A., Gomez, J. Optimization methods applied to renewable and sustainable energy: A review. *Renew Sustain Energy Rev*, 2011, 15, 1753–66. https://doi.Org/10.1016/j. rser.2010.12.008

Bataille, C., Waisman, H., Colombier, M., Segafredo, L., Williams, J. The Deep Decarbonization Pathways Project (DDPP): Insights and emerging issues. *Clim Policy*, 2016. 16, S1–S6.

Brew-Hammond, A. Energy: The Missing Millennium Development Goal. In: Toth, F.L., editor. *Energy dev.*, 2012, Dordrecht: Springer, p. 35–43.

Cherp, A., Jewell, J. The concept of energy security: Beyond the four As. *Energy Pol*, 2014, 75, 415–21. https://doi.Org/10.1016/j.enpol.2014.09.005

Cieplinski, A., D'Alessandro, S., Marghella, F. Assessing the renewable energy policy paradox: A scenario analysis for the Italian electricity market, *Renew Sustain Energy Rev*, 2021, 142, 110838.

Curtin, C., McInerney, B., Gallachóir, Ó., Hickey, C., Deane, P., Deeney, P. Quantifying stranding risk for fossil fuel assets and implications for renewable energy investment: A review of the literature, *Renew Sustain Energy Rev*, 2019, 116, 109402.

Demirbas, M.F., Balat, M., Balat, H. Potential contribution of biomass to the sustainable energy development. *Energy Convers Manag*, 2009, 50, 1746–60. https://doi.Org/10.1016/j.enconman.2009.03.013

Diesendorf, M., Elliston, B. The feasibility of 100% renewable electricity systems: A response to critics, *Renew Sustain Energy Rev*, 2018, 93, 318–30.

Diezmartínez, C.V. Clean energy transition in Mexico: policy recommendations for the deployment of energy storage technologies, *Renew Sustain Energy Rev*, 2021, 135, 110407.

Dincer, I. Renewable energy and sustainable development: A crucial review. *Renew Sustain Energy Rev*, 2000, 4, 157–75. https://doi.org/10.1016/S1364-0321 (99)00011-8

Evans, A., Strezov, V., Evans, T.J. Assessment of sustainability indicators for renewable energy technologies. *Renew Sustain Energy Rev*, 2009, 13, 1082–8. https://doi.Org/10.1016/j.rser.2008.03.008

Fuso Nerini, R., Tomei, J., To, L.S., Bisaga, L., Parikh, R., Black, M., Borrion, A., Spataru, C., Castan Broto, V., Anandarajah, G. Mapping synergies and trade-offs between energy and the Sustainable Development Goals. *Nat Energy*, 2018, 3, 10–15.

Gunnarsdottir, I., Davidsdottir, B., Worrell, E., Sigurgeirsdottir, S. Sustainable energy development: History of the concept and emerging themes. *Renew Sustain Energy Rev*, 2021, 141, 110770, https://doi.org/10.1016/j.rser.2021.110770

Hashim, H., Ho, W.S. Renewable energy policies and initiatives for a sustainable energy future in Malaysia. *Renew Sustain Energy Rev*, 2011. https://doi.org/ 10.1016/j.rser.2011.07.073

Hossein Ghorashi, A., Hadi Maranlou, H. Essential infrastructures and relevant policies for renewable energy developments in oil-rich developing countries: Case of Iran. *Renew Sustain Energy Rev*, 2021, 141, 110839, https://doi.org/10.1016/j.rser.2021.110839

Hu, J., Harmsen, R., Crijns-Graus, W., Worrell, E., van den Broek, M. Identifying barriers to large-scale integration of variable renewable electricity into the electricity market: A literature review of market design. *Renew Sustain Energy Rev*, 2018, 81, 2181–95.

Huang, B.N., Hwang, M.J., Yang, C.W. Causal relationship between energy consumption and GDP growth revisited: A dynamic panel data approach. *Ecol Econ Earn* 2008 67, 41–54.

Hussain, A., Arif, S.M., Aslam, M. Emerging renewable and sustainable energy technologies: State of the art. *Renew Sustain Energy Rev*, 2017, 71, 12–28.

International Atomic Energy Agency, United Nations Department of Economic and Social Affairs, International Energy Agency, Eurostat, European Environment Agency. *Energy indicators for sustainable development: Guidelines and methodologies*, 2005. Vienna.

International Energy Agency, OECD. Energy technology perspectives 2017 – catalysing energy technology transformations, 2017. https://doi.org/10.1787/ energy_tech-2014-en. Paris.

Jefferson, M. Energy Policies for Sustainable Development. In: *World Energy Assessment: Energy and the Challenge of Sustainability*, 2000, UNDP: New York, pp. 415–47.

Jenkins, K., McCauley, D., Forman, A. Energy justice: A policy approach. *Energy Pol*, 2017, 205, 631–34.

Kim, C. A review of the deployment programs, impact, and barriers of renewable energy policies in Korea, *Renew Sustain Energy Rev*, 2021, 144, 110870.

Lund, H. Renewable energy strategies for sustainable development. *Energy*, 2007, 32, 912–19. https://doi.Org/10.1016/j.energy.2006.10.017

Mao, G., Liu, X., Du, H., Zuo, J., Wang, L. Way forward for alternative energy research: A bibliometric analysis during 1994–2013. *Renew Sustain Energy Rev*, 2015, 48, 276–86.

Martinot, E., Chaurey, A., Lew, D., Moreira, J.R., Wamukonya, N. Renewable energy markets in developing countries. *Annu Rev Energy Environ*, 2002, 27, 309–48. https://doi.org/10.1146/annurev.energy.27.122001.083444

Modi, V., McDade, S., Lallement, D., Saghir, J. *Energy Services for the Millennium Development Goals*, 2005, UNDP: New York.

Najam, A., Cleveland, C.J. Energy and Sustainable Development at Global Environmental Summits: An Evolving Agenda. *Environ Dev Sustain*, 2003, 5, 117–38.

Narula, K., Reddy, B.S. Three blind men and an elephant: the case of energy indices to measure energy security and energy sustainability. *Energy*, 2015. 80, 148–58. https://doi.Org/10.1016/j.energy.2014.ll.055

Nerini, F.F., Tomei, J., To, L.S., Bisaga, I., Parikh, P., Black, M. Mapping synergies and trade-offs between energy and the sustainable development goals. *Nat Energy*, 2018, 3, 10–5. https://doi.org/10.1038/s41560-017-0036-5

Olabi, A.G., Abdelkareem, M.A. Renewable energy and climate change, *Renew Sustain Energy Rev*, 2022, 158, 112111, https://doi.org/10.1016/j.rser.2022.112111.

Owen, A.D. Renewable energy: Externality costs as market barriers, *Energy Pol*, 2006, 34, 632–42.

Rosenthal, J., Quinn, A., Grieshop, A.P., Pillarisetti, A., Glass, R.I. Clean cooking and the SDGs: Integrated analytical approaches to guide energy interventions for health and environment goals. *Energy Sustain Dev*, 2018, 42, 152–59.

Sen, S., Ganguly, S. Opportunities, barriers and issues with renewable energy development – A discussion, *Renew Sustain Energy Rev*, 2017, 69, 1170–81.

Sorrell, S., Dimitropoulos, J., Sommerville, M. Empirical estimates of the direct rebound effect: a review. *Energy Pol*, 2009, 37,1356–71. https://doi.org/10.1016/j.enpol.2008.11.026

Sovacool, B.K.; Heffron, R.J.; McCauley, D.; Goldthau, A. Energy decisions reframed as justice and ethical concerns. *Nat Energy*, 2016, 1,16024.

Spalding-Fecher, R., Winkler, H., Mwakasonda, S. Energy and the world summit on sustainable development: What next? *Energy Pol*, 2005, 33, 99–112. https://doi. org/10.1016/S0301-4215(03)00203-9

Spittler, N., Gladkykh, G., Diemer, A., Davidsdottir, B. Understanding the Current Energy Paradigm and Energy System Models for More Sustainable Energy System Development. *Energies*, 2019, 12(8), 1584. https://doi.org/10.3390/en12081584

Ullah, Z., Elkadeem, M.R., Kotb, K.M., Taha, I.B.M., Wang, S. Multi-criteria decision-making model for optimal planning of on/off grid hybrid solar, wind, hydro, biomass clean electricity supply. *Renew Energy*, 2021, 179, 885–910.

UN Commission on Sustainable Development. *Report on the Ninth Session*, 2011, United Nations: New York.

United Nations Department of Economic and Social Affairs. *Indicators of Sustainable Development: Guidelines and Methodologies*, 3rd ed. 2007, New York.

United Nations Development Programme, United Nations Department of Economic and Social Affairs, World Energy Council. *World Energy Assessment: Energy and the Challenge of Sustainability*, 2000, New York.

United Nations General Assembly. *Transforming Our World: The 2030 Agenda for Sustainable Development*, 2015, New York.

United Nations. *Adoption of the Paris Agreement*. Conference of the Parties: Twenty-First Session; United Nations, 2015, New York.

United Nations. Department of Public Information. The Millennium Development Goals, List of Millennium Development Goals, 2000, United Nations: New York.

United Nations. *Johannesburg Declaration on Sustainable Development*; United Nations: Johannesburg, South Africa, 2002, pp. 24–7.

United Nations. *Sustainable Development Goals. About Sustain Dev Goals*, 2022. Available online: www.un.org/sustainabledevelopment/sustainable-development-goals/ (accessed on 2 January 2022).

United Nations. *Sustainable Development. Agenda 21*. 1992, Rio de Janeiro: United Nations.

Wang, S., Li, G., Fang, C. Urbanization, economic growth, energy consumption, and CO2 emissions: empirical evidence from countries with different income levels, *Renew Sustain Energy Rev*, 2018, 81, 2144–2159.

Wichaisri, S., Sopadang, A. Trends and future directions in sustainable development. *Sustain*, 2018, 26, 1–17. https://doi.org/10.1002/sd.1687

World Commission on Environment and Development. *Our Common Future*, 1987, New York.

World Energy Council. *World Energy Trilemma Index 2019*. 2019. London.

Zhang, P., Yanli, Y., Jin, S., Yonghong, Z., Lisheng, W., Xinrong, L. Opportunities and challenges for renewable energy policy in China. *Renew Sustain Energy Rev*, 2009, 13, 439–49. https://doi.Org/10.1016/j.rser.2007.ll.005

Zhu, Z., Xu, Z. The rational design of biomass-derived carbon materials towards next-generation energy storage: a review, *Renew Sustain Energy Rev*, 2020, 134, 110308.

2 Multi-Criteria Decision Aiding and the Governance of Sustainability

2.1 DECISION MAKERS AND STAKEHOLDERS IN THE ENERGY SECTOR

Development of the energy sector requires various types of actors' collaboration to decide on planning and implementing energy initiatives. The plans and approval of the investments in the energy sector depend on the decision makers. The decisions in the energy sector include large-scale projects, and with this comes the risk that many different stakeholders can be affected. Although various stakeholders do not make legal decisions, they have concerns and interests in the sector and are affected by such decisions. Opinions about the relevance of the project and possible benefits, costs, and consequences can significantly vary among the stakeholders. This is due to different points of view, levels of information and knowledge, assessment of personal or societal costs, diverse concerns, political affiliation, and so forth (Geels et al., 2018; Wolsink, 2018; Donald et al., 2021). The technical information and expertise of stakeholders can support the making of decisions, while clarification of stakeholders' needs can help to reflect them in developing the sector. The decision makers in the energy sector can be organized into several main categories and are presented in Figure 2.1.

Also, the stakeholders can be categorized into several groups; these groups are presented in Figure 2.2.

The development of energy projects depends on many different organizations and stakeholders. Many of them do not directly impact project planning and implementation, but can have important information or knowledge, or can influence acceptability of the project in society. Effective stakeholder participation in the energy sector planning allows for achieving the best possible results where the needs of different stakeholders are considered. The involvement of stakeholders in the decision making process gives benefits through knowledge sharing, communication, and collaboration. Therefore, it can be stated that the development of energy sector in a sustainable way depends on active stakeholders' participation in decision making and the flow of information between decision makers and stakeholders (Boudet, 2019; Drozdz et al., 2021).

DOI: 10.1201/9781003327196-2

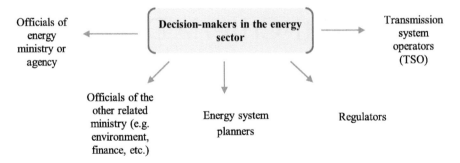

FIGURE 2.1   The main groups of decision makers in the energy sector.

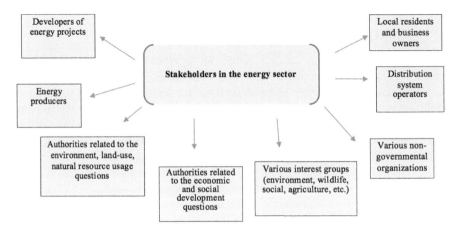

FIGURE 2.2   The main groups of stakeholders in the energy sector.

However, it should be noted that a lot of barriers can be mentioned for an effective and trustful partnership between decision makers and stakeholders, especially citizens (Massay et al., 2018). Examples of the barriers can be mentioned:

- Lack of communication between local authorities and citizens and effective mechanisms to reach them;
- Lack of transparency in energy policy planning, policy implementation, and sector development;
- Lack of knowledge about advanced energy technologies;
- Low public perception and acceptance of new innovative energy technologies;
- Stakeholder engagement is not considered as an important part in the energy policy decision making;
- Not enough attention is paid to social and environmental aspects;
- Lack of motivation and interest by decision makers to engage stakeholders;
- Lack of resources to engage stakeholders;

- Not all groups of stakeholders are involved in the process;
- The balance between different groups of stakeholders is not correct, and so forth.

The engagement of various stakeholders can bring multiple benefits to local communities and municipalities, and mutual benefit between different groups of stakeholders. Attention should especially be paid to involving local communities in the development of new energy projects. This allows for democratizing the decision making processes in the energy sector and finding consensus on implemented strategies and measures. Usually, the engagement of residents is encouraged with the goals oriented towards energy consumers – goals such as savings in energy costs, improvements in quality of daily life, new local job places, and so forth (Athavale et al., 2021). These goals allow not only involving local people in the decision making, but also to reduce the NIMBY (not in my backyard) phenomenon and raise awareness regarding new innovative energy technologies and efficient energy usage. Inclusive policy making can help in understanding why opposition occurs regarding transition to clean energy measures and in reflecting the concerns in shaping the energy sector. Also, it is good to inform the stakeholders regarding the policy and measures and thereby address potential opposition.

2.2   PARTICIPATION AND DELIBERATION AS A TOOL FOR INCLUSIVE POLICY MAKING

Decisions in the energy sector are complex. They cannot be solved effectively by one institution or interest group. Not including all the parties in project planning that will be affected will make the project less successful. Ansell and Gash (2008) defined collaborative governance as process whereby one or more public organizations directly involves non-state stakeholders in a decision making process that is official and oriented to the consensus and seeks to implement public policy or manage public programs. The two key elements can be identified for successful collaborative governance process: cross-sector collaboration and engagement of various stakeholders in decision making (Fischer et al., 2020). Cross-sector collaboration shows the importance and connection between sharing knowledge, competencies, and resources by organizations across different sectors (Bryson et al., 2006). The engagement of various stakeholders, including residents, is the other key component of collaborative governance, and it is extremely important to the success of the whole process of energy project implementation. Stakeholders' participation and empowerment of stakeholders in decision making can aid as a driver for transparent decision making, information sharing and learning, increasing acceptance of energy initiatives, and reaching more acceptable decisions (Gustafsson et al., 2015).

Successful stakeholder participation in developing the energy sector requires the active engagement of decision makers and stakeholders. Four main steps can be singled out for successful stakeholder involvement (Figure 2.3).

(1) Stakeholder identification step. The stakeholders related to the project and the proper balance of different groups of stakeholders are determined in the stakeholder

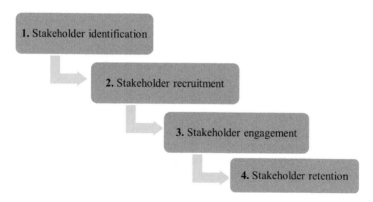

FIGURE 2.3   The main steps for the stakeholder involvement.

identification step. This should be done at the beginning of the project to ensure that stakeholders are involved throughout all the steps of the project. In order to identify all stakeholders who may affect or be affected by the project, several questions can be raised:

- Who can be affected by the project being implemented?
- Who can benefit from the project being implemented?
- Who can contribute to the project implementation?

Also at this stage, it would be useful to identify a leader to facilitate discussions among different groups of stakeholders. Such a leader should have enough technical expertise and influence on the project to effectively work as a process facilitator, supporting multi-stakeholder dialogue and consensus building. It can be stated, that the leadership is very important for gathering different parties to the discussion table (Ansell and Gash, 2008), setting and complying with the rules, building trust, managing discussions, exploring mutual benefits (Vangen and Huxham, 2003), and empowering and representing less active or weaker stakeholders (Ozawa, 1993).

(2) Stakeholder recruitment step. It is very important not only to involve all stakeholders in the process, but also to create the appropriately sized group. Too small a group can overlook important concerns, and a too large group can deal with too many discussions in different directions and ultimately become ineffective. Therefore, it is necessary to select those stakeholders who may have the biggest influence on the project. The main attention should be paid to those stakeholders who have negative opinions regarding the project and its implementation, because they may have a high impact on the final project outcomes. The involvement of various stakeholders gives the possibility to reflect stakeholders needs, reduce resistance regarding the project, and boost the satisfaction with project results.

(3) Stakeholder engagement step. Open and transparent communication when engaging stakeholders is necessary to create a sense of inclusion. The stakeholders can be engaged in all steps of project processes through sharing relevant information

and knowledge, consulting on issues raised in different project planning and implementation steps, providing feedback, offering insights, disseminating outcomes of the project, or increasing the acceptability of the project in the local community.

(4) Stakeholder retention step. The relationships with stakeholders should be sustained, not only throughout all the project implementation steps, but also when the project is already finished. This allows for a comprehensive long-term evaluation of the implemented energy project and provides guidelines for the development of successful future energy projects.

The involvement of various stakeholders in solving energy-related issues, especially regarding the transition to clean energy, allows for dealing with social conflicts. An example can be provided by Sovacool et al. (2022), where 130 cases of opposition to different energy infrastructure projects were analyzed. The results showed that renewable and nuclear energy are of the greatest public concern in many cases. Meanwhile, international pressure encourages actions, shapes the actors involved, and affects the results achieved. Also, it was found that the energy transition policy in carbon-intensive regions differs considerably. The cases of community mobilization against energy projects shows how important is to pay attention, not only to economical or technological aspects of the project and its outcomes, but also to give close attention to social aspects and questions raised in the community regarding the project. Therefore, a smooth energy transition is possible only with collaboration between decision makers and stakeholders. Only in that way it is possible to reflect the needs of society and manage societal response.

2.3 MULTI-CRITERIA DECISION MAKING AS A TOOL TO DEAL WITH COMPROMISES

Energy planning issues have become very important in decision making since the early 1980s, with the aim to reduce energy prices and to solve energy scarcity problems. The growing negative impact on the environment and concerns regarding climate change has updated the aspects under consideration of decision makers, adding environmental issues to the energy planning programs and decision making. As well as the growing concern regarding the improvements in social policy, issues of social welfare in development of energy sector arose. The contradiction of goals is obvious when solving energy issues: what is best from an environmental point of view may be unacceptable from an economic perspective; the most cost-effective solution may be socially unacceptable, and so forth. Reasonably and adequately selected mathematical models allow for combining conflicting energy policy aspects and different needs of various stakeholders according to the criteria selected. Therefore, it can be stated, that the usage of multi-criteria techniques allows for dealing with compromises.

Sustainability assessment is actively used today as one of the main preventive instrument to reduce impact on the environment. Numerous systems, methodologies, and assessment instruments have been applied to energy sustainability studies. Many rely on different multi-criteria analysis techniques. The application of multi-criteria decision making (MCDM) instruments to solving energy questions started to

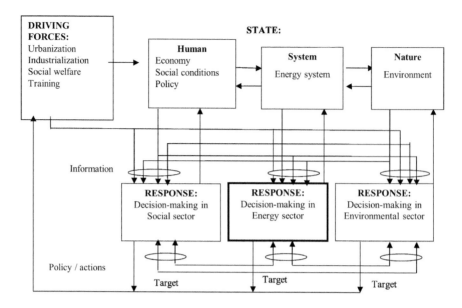

FIGURE 2.4   The linkages between sustainability dimensions and decision making.

be popular in the last decade. Currently, multi-criteria analysis is widely applied in studies related to sustainability measurement (Campos-Guzman et al., 2019; Deng et al., 2022), energy policy, or national and regional planning (Witt and Klumpp, 2021, Villacreses et al, 2022), project and site selection (Ridha et al., 2021; Shao et al., 2020; Zolfani et al., 2022), technologies assessment (Bohra and Anvari-Moghaddam, 2021; Abdul et al., 2022), and so forth.

MCDM techniques are particularly useful when dealing with prioritization problems, when decisions should be made based on several conflicting indicators or when indicators selected are competing for their importance. Siksnelyte et al. (2018) modified the improved Driving Force, State and Response (DSR-HNS) policymaking framework presented by Meyar-Naimi and Vaez-Zadeh (2012) to show linkages between sustainability dimensions and decision making in the energy sector (Figure 2.4). These linkages indicate the necessity to perform decision making processes in sustainable ways with compromise result.

The versatility and wide range of multi-criteria methods application for solving sustainable energy issues is revealed in Figure 2.5, showing keywords from the Web of Science database on a combination of topics: "sustainable energy", "energy sustainability", "multi-criteria decision making", "multi-criteria decision analysis" (Web of Science database, 2022). It should be noted that the use of terms in the scientific literature is not uniform. Examples can be multi-criteria decision making; multi-criteria decision-making; multi criteria decision making; multi criteria decision-making; multicriteria decision making, multicriteria decision-making. It can increase the risk, that, due to the inappropriate use of terms, important studies cannot get as much attention as they deserve.

FIGURE 2.5 Variety and popularity of keywords in studies dealing with sustainable energy issues.

MCDM approaches allow for performing integrated assessment considering selected indicators in a comprehensive manner, measuring interrelations among them and different degrees of significance. Techniques used in multi-criteria analysis can be grouped into two categories: multi-objective decision making (MODM) and multi-attribute decision making (MADM) techniques. Both mentioned approaches distinguish with similar characteristics. MODM techniques handle questions where the decision space is continuous with the aim of finding the most preferable solution within predefined boundaries. In contrast, MADM methods focus on problems with discrete decision spaces where a set of possible solutions has already been established (Triantaphyllou et al., 1998; Baumann et al. 2019).

MCDM procedure can be subdivided into two basic steps (Construction and Exploitation), which are overlapping and cannot be separated from each other (Guitouni and Martel, 1998). The first step is Construction, where the main tasks are:

- the goal, scope and possible alternatives should be defined;
- the involvement of different stakeholders should be considered;
- criteria for the evaluation should be identified and selected.

In the Exploitation step, the following tasks can be singled out:

- performance of the criteria selected should be measured;
- the most suitable technique should be selected;

- criteria aggregation procedure should be performed;
- weights and weighting schemes should be determined;
- the results should be compared.

The decision making process and selection of appropriate instruments for decision making are very important issues, having clear relationships with policy targets implementation. The application of MCDM techniques can help to deal with contradictory questions, to indicate the best alternative based on the criteria selected (target values), and to combine different policies. Also, multi-criteria analysis can be applied not only when making decisions, but also when creating objectives, and searching for the measures to attain them. Due to the wide applicability of multi-criteria techniques, they are increasingly being used to deal with energy policy issues.

2.4   SELECTION OF INDICATORS

Many evaluation systems have been developed to assess sustainability of different energy issues. Depending on what are assessed (e.g., the technologies, the sector as a whole, the measures implemented, etc.), the indicators and the dimensions reflected vary. Mostly, the traditional sustainable development approach is applied for the sustainability assessment studies, and the selected indicators reflect three dimensions of sustainability (economical, social, and environmental). Selection of indicators should play a key role in the decision making. The indicators selected should not only reflect the different needs of various stakeholders or different dimensions of sustainability, but the fundamental requirements for indicators' selection should also be met. The main requirements for the indicators' selection can be singled out; they are provided in Table 2.1 (Hirschberg, et al., 2008).

For the assessment and comparison of several countries, it is important to create a set of indicators that not only entirely reflect and evaluate the problems, but also find

TABLE 2.1
The Fundamental Requirements for the Indicators Selection

Requirement	Explanation
Representative	Set includes essential characteristics of the problem analyzed, indicators are clear in value and content.
Understandable and concrete	All indicators are measurable, no double counting and direct as possible.
Sensitive	Indicators are responsive to changes and easy comparable under the study.
Verifiable	Computation of indicators is quite easy, data are public available, measurement is justified methodologically and repeatable.

Source: Created by authors.

FIGURE 2.6   Steps of indicators set creation.

all the comparable data for the assessment. Data availability is one of the aspects that often limits scientists to analyze many countries in one study. Also, data availability is a condition on which depends the possibility to repeat the evaluation in the future or compare and monitor the results.

The properly selected indicators can be useful tools for the decision making related to the questions of energy policy and sustainable development for policy makers. Various indicators allow for systematizing and describing statistical data that reflect the state of economy, the environment, and social welfare. Thoroughly selected indicators can also serve to monitor progress and evaluate policy measures and their impacts to objectives of sustainable development. The creation of indicators sets can be divided into four main steps (Figure 2.6).

In the first step, a set of indicators that reflect economical, social, environmental aspects of the problem are reviewed and analyzed. In the second step, indicators are divided into groups. At the start, a large number of indicators of each category may be defined, for example: descriptive, comparative, normalized, structural, stress, decomposition, consequence, causal, and physical indicators (Patlitzianas, 2008). When selecting the indicators, close attention should be paid to their simplicity, comparability, technical-scientific adequacy, and compliance with the standard. In the third step, the indicators are prioritized taking into account the time horizon and the impact on the environment. The fourth step provides a possibility to supplement and correct the set created, and a final set is created (set (verification of reliability, duplication checking, etc.).

In order to achieve sustainable development, the decision making should be performed in a sustainable way. The Bellagio Sustainability Assessment and Measurement Principles (STAMP) are very helpful tools for setting the indicators and in the process of the evaluation (Figure 2.7). These guidelines have been created by the International Institute for Sustainable Development and the Organisation for Economic Co-operation and Development (2009) and they provide common guidelines on which the following evaluation performed, fully reflecting sustainability issues (Pinter et al., 2012).

(1) Guiding Vision principle seeks to ensure that the goal of well-being for future generations will be considered in the assessments.
(2) Essential Considerations principle stress the importance of considering all dimensions of sustainability as a whole and interactions between them, dynamics and current trends as well as drivers, risk, activities, and uncertainties of changes. Also, the adequacy of governance mechanisms and implications for decision making should be considered.

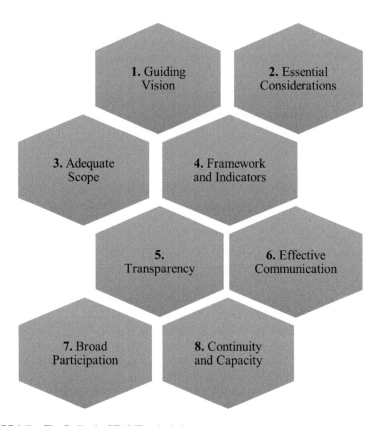

FIGURE 2.7   The Bellagio STAMP principles.

(3) Adequate Scope principle highlights the importance to assess proper time horizon for the consideration of short- and long-term effects of human activities and policy decisions. Also, the appropriate geographical scope, which ranges from local to global, should be taken.

(4) Framework and Indicators principle defines that assessment should be based on framework, which includes all indicators reflecting all the domains under consideration. The data for the evaluation should be the newest and most reliable, the methods for the assessment should be standardized, allowing to make projections and comparisons, determine trends, and create possible scenarios.

(5) Transparency Principle requires ensuring public data and results availability to provide clear explanation of applied assumptions and choices, to reveal data sources used, methods applied, foundation sources, and potential conflicts of interest.

(6) Effective Communication principle seeks to ensure the broad audience attraction. In order to implement this, the usage of clear and plain language should be ensured. Also, the information should be presented in an objective way, and visual tools to aid interpretation of the assessment should be used.

(7) Broad Participation principle stresses the importance of reflecting public opinion in order to strength the relevance of the assessment.

(8) Continuity and Capacity principle requires that the assessment would be repeated and the opportunity to adapt the assessment to the changing conditions would be provided.

The Bellagio STAMP guide the whole process of the sustainability assessment, including the process of indicators selection, interpretation and communication of the results. The guidelines are designed to help decision makers assess societal progress, considering policy options or supporting changes in society. All eight principles should be used in the evaluations, because they are interrelated and intended to be applied as a set. Following these principles, the content, process, scope and impact of the evaluation is supported. The Bellagio STAMP supports the content of the assessment, by guiding what questions should be answered, supports the process of the assessment by guiding the way in which the evaluation should be performed, helps to range the assessment across time and geography aspects and stress the importance of the assessment.

In the scientific literature, there are attempts to apply these principles to the decision making. For example, Bartke and Schwarze (2015) analyzed the role of the sustainability assessment tools in land-use management. The study sought to ascertain how various tools reflect the Bellagio STAMP. Arulnathan et al. (2020) applied the Bellagio STAMP for the assessment of decision making tools for farms. Adewumi et al. (2019) overviewed different neighborhood sustainability assessment frameworks and determined areas for improvement using the Bellagio STAMP. Gunnarsdottir et al. (2020) created a framework based on the Bellagio STAMP and evaluated indicators set for sustainable energy development. Siksnelyte-Butkiene et al. (2022) applied these principles for the creation of framework and assessment procedures for energy poverty assessment at country level. The methodological recommendations for the energy studies following the Bellagio STAMP are provided in Table 2.2.

TABLE 2.2
The Methodological Recommendations for the Energy Studies Following the Bellagio STAMP Principles

Principle	Implementation	Interpretation
Guiding vision	• Focus on the sustainable development goals.	• The selected indicators reflect needs not only present but also future generations.
Essential Considerations	• Following the concept of sustainability. • Evaluation of dynamics and interaction between current trends and drivers.	• Indicators reflecting all dimensions of sustainability are involved in the assessment. • The set created allows for evaluation of the current or past situation in the energy sector and the assessment can be repeated in the future.

(continued)

TABLE 2.2 (Continued)
The Methodological Recommendations for the Energy Studies Following the Bellagio STAMP Principles

Principle	Implementation	Interpretation
Adequate Scope	• Short- and long-term effects. • The alternatives to be evaluated are properly selected.	• The indicators set allows for determining both short- and long-term impact of the current economic, environmental, energy and social policy decisions and human activity. • The selected alternatives for the assessment should be comparable and belong to the same category. For example, for the countries comparison, all countries under analysis should belong to the same economical and political region, the economic and political situation of the countries can be possible to compare in different types of assessments.
Framework and Indicators	• Conceptual framework. • Standardized measurement techniques and indicators comparability.	• The framework reflects the main aspects of the problem analyzed and allows to monitor the progress achieved. • In order to track the progress achieved, all selected indicators should be standardized and comparable. The assessment techniques selected should be justified scientifically and recognized as suitable tools for that type of research.
Transparency	• Data availability. • Indicators justification. • Data sources and evaluation tools are clearly indicated.	• The data used should be publicly available. • The indicators set should be developed after a comprehensive literature analysis, and the weights of indicators should be justified. • All data and literature sources used and methods should be clearly indicated.
Effective Communication	• Objectivity of presentation. • Usage of visual instruments and graphical interpretation.	• The assessment should be carried out by application of scientific techniques, the interpretation of results should follow the objectivity requirements. • Visual tools should be used to simplify the interpretation of the results and to attain the wide audience as possible.
Broad Participation	• Experts participation. • Stakeholders involvement.	• Experts' participation is desirable, for example to determine the indicators weights. • Various methods to involve stakeholders or the public in the assessment are desirable, especially to reflect social dimension.

TABLE 2.2 (Continued)
The Methodological Recommendations for the Energy Studies Following the Bellagio STAMP Principles

Principle	Implementation	Interpretation
Continuity and Capacity	• Assessment repeatability.	• The data availability, the clearance of the assessment methodology, the justification of indicators and their weights make the assessment repeatable and progress can be tracked easily.

Source: Created by authors.

2.5 COMPARATIVE EVALUATION OF THE MOST POPULAR MCDM TECHNIQUES

Many different MCDM techniques have been applied to solve various questions in the energy sector. However, some of them can be singled out as more popular and have been applied more commonly in sustainable energy studies. The most commonly applied MCDM techniques to solve sustainable energy development issues are discussed in this section.

Siksnelyte et al. (2018) performed a systematic literature review and overview studies, which used multi-criteria analysis for the sustainable decision making in the energy sector. According to the results, the most popular MCDM techniques are: Analytic Hierarchy Process (AHP) and Technique for Order Preference by Similarity to Ideal Solutions (TOPSIS). Kaya et al. (2018) overviewed the studies for energy policy and decision making issues. The same AHP and TOPSIS methods as the most applied MCDM tools identified. Siksnelyte-Butkiene et al. (2020) performed literature analysis of the MCDM techniques for the assessment of household-level renewable energy technologies and identified the same methods as most applied in previously mentioned studies.

The most applied MCDM approaches can be classified in decision making as pairwise comparison based techniques (AHP and Analytic Network Process (ANP)), distance-based techniques (TOPSIS and Multi-Criteria Optimization and Compromise Solution (VIKOR)), outranking techniques (Preference Ranking Organization Method for Enriching Evaluation (PROMETHEE) and Elimination and Choice Transcribing Reality (ELECTRE)), and other methods (Kaya et al., 2018). The AHP and ANP methods are usually applied to calculate evaluation criteria weights. The methods are a useful tool, but have difficulties confirming consistency of pairwise comparisons. Also, the methods can be used for the assessment of alternatives. TOPSIS and VIKOR techniques are used to evaluate the alternatives according to their distances to ideal solutions. The PROMETHEE and ELECTRE techniques are used to assess the alternatives by means of outranking relations between them. It necessary to highlight that each technique has its own advantages and disadvantages. The selection of the most suitable technique should be identified by considering all

TABLE 2.3
Advantages and Disadvantages of MCDM Methods

		AHP	TOPSIS	VIKOR	ELECTRE	PROMETHEE
Advantages	Easy computation process	x	x	x		
	Low time costs	x	x	x		
	Non-compensatory				x	x
	Comprehensible logic of calculations		x	x		
	Robust to outliers			x		x
Disadvantages	Additional analysis is required to verify the results	x			x	
	Subjective assumptions are required	x			x	x
	Due to sophisticated calculations is suitable only for experts				x	x
	High time costs				x	x
	Not suitable, when values differ very strongly		x			

Source: Created by authors.

the advantages and disadvantages of the methods, the suitability for the research due to available data, experience of decision makers, the accuracy of the desired result, and possible time cost.

Table 2.3 provides comparative evaluation of the most popular MCDM methods that were applied for energy-related decision making issues.

The AHP technique is based on a pairwise comparison scale and was introduced by Saaty et al. (Saaty et al., 1980). This technique is the most popular MCDM method used to solve energy-related questions. The method can be easily applied to solve many energy-related problems, but usually is used for weighting. The mathematical form does not include sophisticated calculations, therefore the results can be obtained quickly and calculations do not require expertise. However, due to subjective assumptions required, the evaluation process can be complicated, when involved with more than one decision maker. Also, additional analysis is required for the verification of the results obtained (Saaty, 2004; Ishizaka and Labib, 2009; Shahroodi et al., 2012; Kumar et al., 2017).

The TOPSIS technique is the second most popular MCDM tool in energy studies. The technique was introduced by Hwang and Yoon (1981) and relies on measuring the distance to the ideal solution (Jato-Espino et al., 2014). As AHP, the TOPSIS is distinguished for fairly simple calculations and rapidly obtained results. The logic of computation is rational, understandable, and expressed in quite simple mathematical form. From the decision maker perspective, it is easy to interpret the results and understand the importance of the criteria selected for the final results. It is necessary to highlight that TOPSIS is based on Euclidean distance, and positive and negative values of criteria do not have differences in calculations. Also, the method is not suitable when significant deviations among values of the alternatives are observed – indicators differ significantly among themselves (Shih et al., 2007; Boran et al., 2009).

The VIKOR method was introduced by Opricovic (1998) and now is widely used in various fields of decision making or is integrated with other techniques (Mardani et al., 2016). The technique seeks to determine the closeness to the positive and the negative ideal solution. But, different from TOPSIS method, the VIKOR takes into account the relative importance of the distances from the positive and the negative ideal solutions (Opricovic and Tzeng, 2007). The method is tolerant to deviations of criteria values among the alternatives. The calculation process is quite simple and the results can be obtained quickly compared with other methods. Despite that, the results could be affected by the normalization procedure and weighting strategy.

Among the most accurate and effective methods are the PROMETHEE methods. The first version was developed in 1986 by Brans et al. (1986). The qualitative and quantitative information as well as the uncertain information can be used in the calculations. Also, PROMETHEE technique is very useful when it is difficult to harmonize the alternatives. In addition, the alternatives that are highly interchangeable can be compared (Wang et al., 2010; Amaral and Costa, 2014; Brans and de Smet, 2016). However, the method has complex mathematical expressions (Kumar et al., 2017; Alinezhad and Khalili, 2019) and requires specific knowledge and experience. Also, subjective assumptions are required and the calculations require a high time cost. In general, the method is intended only for professionals in such research.

The first ELECTRE method was presented by Roy in 1968 (Roy, 1968). ELECTRE needs determination of the concordance and discordance indices, and this process involves lengthy computations. The ELECTRE method can be distinguished as requiring high time costs and subjected assumptions, because the decision maker has to select threshold values for the calculation of concordance and discordance indices (Karande and Chakraborty, 2012). It is also recognized that for verification of the results additional analysis is required.

REFERENCES

Abdul, D., Jiang, W.Q., Tanveer, A. Prioritization of renewable energy source for electricity generation through AHP-VIKOR integrated methodology. *Renew Energy*, 2022, 184, 1018–32.

Adewumi, A.S., Onyango, V., Moyo, D., Al Waer, H. A review of selected neighbourhood sustainability assessment frameworks using the Bellagio STAMP. *Int J Build Pathol Adapt*, 2019, 37(1), 108–18.

Alinezhad A., Khalili, J. New Methods and Applications in Multiple Attribute Decision Making (MADM). *International Series in Operations Research & Management Science*, 227, 2019, Switzerland, 203 p., ISSN 2214-7934.

Amaral, T.M., Costa, A.P. Improving decision making and management of hospital resources: an application of the PROMETHEE II method in an Emergency Department. *Oper Res Health Care*, 2014, 3(1), 1–6.

Ansell, C., Gash, A. Collaborative governance in theory and practice. *J Public Adm Res Theory*, 2008, 18, 543–71.

Arulnathan, V., Heidari, M.D., Doyon, M., Li, E., Pelletier, N., Farm-level decision support tools: A review of methodological choices and their consistency with principles of sustainability assessment. *J Clean Prod*, 2020, 256, 120410.

Athavale, S., Barreto-Gomez, L., Brandl, G., Kassai, A., Köstler, R. Guidelines for local authorities on stakeholder engagement. 2021, 89. Available at: www.excite-project.eu/uploads/9/8/8/4/9884716/d4.1_guidelines_for_local_authorities_on_stakeholder_engagement.pdf

Bartke, S., Schwarze, R. No perfect tools: Trade-offs of sustainability principles and user requirements in designing support tools for land-use decisions between greenfields and brownfields. *J Environ Manage*, 153, 2015, 11–24.

Baumann, M., Weil, M., Peters, J.F., Chibeles-Martins, N., Moniz, A.B. A review of multi-criteria decision making approaches for evaluating energy storage systems for grid applications. *Renew Sust Energ Rev*, 2019, 107, 516–34.

Bohra, S.S., Anvari-Moghaddam, A. A comprehensive review on applications of multicriteria decision-making methods in power and energy systems. *Int J Energy Res*, 2022, 46(4), 4088–118.

Boran, F.E., Genc, S., Kurt, M., Akay, D. A multi-criteria intuitionistic fuzzy group decision making for supplier selection with TOPSIS method. *Expert Syst Appl*, 2009, 36, 11363–11368.

Boudet, H.S. Public perceptions of and responses to new energy technologies. *Nat Energy*, 2019, 4(6), 446–55.

Brans, J.P., De Smet, Y. PROMETHEE Methods. In: Greco S., Ehrgott M., Figueira J. (eds) *Multiple Criteria Decision Analysis. International Series in Operations Research & Management Science*, 2016, 233, 187–219.

Bryson, J.M., Crosby, B.C., Stone, M.M. The Design and Implementation of Cross-Sector Collaborations: Propositions from the Literature. *Public Adm Rev*, 2006, 66, 44–55.

Campos-Guzman, V., Garcia-Cascales, M.S., Espinosa, N.A. Urbina Life Cycle Analysis with Multi-Criteria Decision Making: A review of approaches for the sustainability evaluation of renewable energy technologies. *Renew Sust Energ Rev*, 2019, 104, 343–66.

Deng, D.Q., Li, C., Zu, Y.F., Liu, L.Y.J., Zhang, J.Y., Wen, S.B. A Systematic Literature Review on Performance Evaluation of Power System from the Perspective of Sustainability. *Front Environ Sci*, 2022, 10, 925332.

Donald, J., Axsen, J., Shaw, K., Robertson, B. Sun, wind or water? Public support for large-scale renewable energy development in Canada. *J Environ Policy Plan*, 2021, 1–19, https://doi.org/10.1080/1523908X.2021.2000375

Drozdz, W., Kinelski, G., Czarnecka, M., Wojcik-Jurkiewicz, M., Marouskova, A., Zych, G. Determinants of Decarbonization – How to Realize Sustainable and Low Carbon Cities? *Energies*, 2021, 14, 2640.

Fischer, J., Alimi, D., Knieling, J., Camara, C. Stakeholder Collaboration in Energy Transition: Experiences from Urban Testbeds in the Baltic Sea Region. *Sustainability*, 2020, 12, 9645.

Geels, F.W., Schwanen, T., Sorrell, S., Jenkins, K., Sovacool, B.K. Reducing energy demand through low carbon innovation: a sociotechnical transitions perspective and thirteen research debates. *Energy Res Soc Sci*, 2018, 40, 23–35.

Guitouni, A., Martel, J-M. Tentative guidelines to help choosing an appropriate MCDA method. *Eur J Oper Res*, 1998, 109(2), 501–21.

Gunnarsdottir, I., Davidsdottir, B., Worrell, E., Sigurgeirsdottir, S. Review of indicators for sustainable energy development. *Renew Sust Energ Rev*, 2020, 133, 110294.

Gustafsson, S., Ivner, J., Palm, J. Management and stakeholder participation in local strategic energy planning – Examples from Sweden. *J Clean Prod* 2015, 98, 201–12.

Hirschberg, S., Bauer, B., Burgherr, P., Dones, R., Schenler, W., Bachmann, T., Gallego-Carrera, D. Final set of sustainability criteria and indicators for assessment of electricity supply options. NEEDS Deliverable D3.2 – RS2b. 2008, 36 p. Available at: www.psi.ch/sites/default/files/import/ta/NeedsEN/RS2bD3.2.pdf

Hwang, C.L., Yoon, K. *Multiple Attributes Decision Making Methods and Applications*. Springer: Berlin, Hedelberg, Germany, 1981; p. 22–51.

International Institute for Sustainable Development; Organisation for Economic Co-operation and Development. BellagioSTAMP: Sustainability Assessment and Measurement Principles. 2009, 6 p. Available at: www.iisd.org/system/files/2021-08/bellagio-stamp-brochure.pdf

Ishizaka, A., Labib, A. Analytic hierarchy process and expert choice: Benefits and limitations. *OR Insight*, 2009, 22(4), 201–20.

Jato-Espino, D., Castillo-Lopez, E., Rodriguez-Hernandez, J., Canteras-Jordana, J.C. A review of application of multi-criteria decision making methods in construction. *Autom Constr*, 2014, 45, 151–62.

Karande, P., Chakraborty, S. Application of multi-objective optimization on the basis of ratio analysis (MOORA) method for materials selection. *Mater Des*, 2012, 37, 317–24.

Kaya, I., Colak, M., Terzi, F. Use of MCDM techniques for energy policy and decision-making problems: A review. *Int J Energy Res*, 2018, 42(7), 2344–72.

Kumar, A., Sah, B., Singh, A.R., Deng, Y., He, X., Kumar, P., Bansal, R.C. A review of multi criteria decision making (MCDM) towards sustainable renewable energy development. *Renew Sust Energ Rev*, 2017, 69, 596–609.

Mardani, A., Zavadskas, E.K., Govindan, K., Senin, A.A., Jusoh, A. VIKOR Technique: A Systematic Review of the State of the Art Literature on Methodologies and Applications. *Sustainability*, 2016, 8, 37.

Massay, B., Verna, P., Khadem, S. Citizen Engagement as a Business Model for Smart Energy Communities. 5th International Symposium on Environment-Friendly Energies and Applications (EFEA), 2018. Available at: https://ieeexplore.ieee.org/document/8617063

Meyar-Naimi, H., Vaez-Zadeh, S. Developing a DSR-HNS Policy Making Framework for Electric Energy Systems. *Energy Policy*, 2012, 42, 616–27.

Opricovic, S. Multicriteria Optimization of Civil Engineering Systems. PhD Thesis, Faculty of Civil Engineering, Belgrade, 1998, 302.

Opricovic, S. Tzeng, G.H. Extended VIKOR method in comparison with outranking methods. *Eur J Oper Res*, 2007, 178, 514–29.

Ozawa, P. Improving citizen participation in environmental decision making: The use of transformative mediator techniques. *Environ Plann C Gov Policy*, 1993, 11, 103–17.

Patlitzianas, K.D., Doukas, H., Kagiannas, A.G., Psarras, J. Sustainable energy policy indicators: Review and recommendations. *Renew Energ*, 2008, 33, 966–73.

Pinter, L., Hardi, P., Martinuzzi, A., Hall, J. Bellagio STAMP: principles for sustainability assessment and measurement. *Ecol Indic*, 2012, 17, 20–28.

Ridha, H.M., Gomes, C., Hizam, H., Ahmadipour, M., Heidari, A.A., Chen, H.L. Multi-objective optimization and multi-criteria decision-making methods for optimal design of standalone photovoltaic system: A comprehensive review. *Renew Sust Energ Rev*, 2021, 135, 110202.

Roy, B. La methode ELECTRE. Revue d'Informatique et. de Recherche Operationelle (RIRO). 1968, 8, p. 57–75.

Saaty, T.L. Decision making – the analytic hierarchy and network processes (AHP/ANP). *J Syst Sci Syst*, 2004, 13(1), 1–35.

Saaty, T.L. *The Analytic Hierarchy Process*; McGraw-Hill: New York, 1980; pp. 11–29.

Shahroodi, K., Keramatpanah, A., Amini, S., Sayyad Haghighi, K. Application of analytical hierarchy process (AHP) technique to evaluate and selecting suppliers in an effective supply chain. *Kuwait Chapter Arab J Bus Manag Rev*, 2012, 1(6), 119–32.

Shao, M., Han, Z.X., Sun, J.W., Xiao, C.S., Zhang, S.L., Zhao, Y.X. A review of multi-criteria decision making applications for renewable energy site selection. *Renew Energy*, 2020, 157, 377–403.

Shih, H.S., Shyur, H.J., Lee, E.S. An extension of TOPSIS for group decision making. *Math Comput Model*, 2007, 45, 801–13.

Siksnelyte, I., Zavadskas, E.K., Streimikiene, D., Sharma, D. An Overview of Multi-Criteria Decision Making Methods in Dealing with Sustainable Energy Development Issues. *Energies*, 2018, 11(10), 2754.

Siksnelyte-Butkiene, I., Streimikiene, D., Balezentis, T. Addressing sustainability issues in transition to carbon-neutral sustainable society with multi-criteria analysis. *Energy*, 2022, 254(2), 124218.

Siksnelyte-Butkiene, I., Zavadskas, E.K., Streimikiene, D. Multi-criteria decision-making (MCDM) for the assessment of renewable energy technologies in a household: a review. *Energies*. 2020, 13(4), 1164.

Sovacool, B.K., Hess, D.J. Cantoni, R., Lee, D., Brisbois, M.C., Walnum, H.J., Dale, R.F., Rygg, B., Korsnes, M., Goswami, A., Kedia, S., Goel, S. Conflicted transitions: Exploring the actors, tactics, and outcomes of social opposition against energy infrastructure. *Glob Environ Res*, 2022, 73, 102473.

Triantaphyllou, E., Shu, B., Nieto Sanchez, S., Ray, T. Multi-criteria decision making: an operations research approach. *J Electr Electron Eng*, 1998, 15, 175–86.

Vangen, S., Huxham, C. Enacting leadership for collaborative advantage: Dilemmas of ideology and pragmatism in the activities of partnership managers. *Br J Manag*, 2003, 14, 61–76.

Villacreses, G., Martinez-Gomez, J., Jijon, D., Cordovez, M, Geolocation of photovoltaic farms using Geographic Information Systems (GIS) with Multiple-criteria decision-making (MCDM) methods: Case of the Ecuadorian energy regulation. *Energy Rep*, 2022, 8, 3526–48.

Wang, M., Lin, S.J., Lo, Y.C. The comparison between MAUT and PROMETHEE. International Conference on Industrial Engineering and Engineering Management (IEEM), 2010, 753–757.

Web of Science database, 2022, available at: www.webofscience.com/wos/woscc/basic-search

Witt, T., Klumpp, M. Multi-Period Multi-Criteria Decision Making under Uncertainty: A Renewable Energy Transition Case from Germany. *Sustainability*, 2021, 13(11), 6300.

Wolsink, M. Social acceptance revisited: Gaps, questionable trends, and an auspicious perspective. *Energy Res Soc Sci*, 2018, 46, 287–95.

Zolfani, S.H., Hasheminasab, H., Torkayesh, A.E., Zavadskas, E.K., Derakhti, A. A Literature Review of MADM Applications for Site Selection Problems – One Decade Review from 2011 to 2020. *Int J Inf Technol Decis Mak*, 2022, 21(1), 7–57.

3 Multi-Criteria Decision Making for Regional and National Planning

3.1 SUSTAINABLE REGIONAL AND NATIONAL ENERGY SYSTEM PLANNING

Sustainable development of the energy sector and low-carbon energy transition are the key challenges for many developed countries and the essential tools to fight against climate change (Sangroya et al., 2020; Raghutla and Chittedi, 2020). Different countries have used various measures to encourage the level of sustainability in the energy sector. However, the transition to low-carbon energy and sustainable functioning of the whole energy sector is not a result of a single instrument, but a complex of initiatives that must be scrupulously planned, implemented in an effective and successful way, and regularly monitored (Frank et al., 2020). The implementation of sustainability goals is incorporated in the political documents of many countries, with ambitious targets for renewable energy development, efficiency improvement, reduction of greenhouse gas (GHG) emissions, and so forth. However, the approach of sustainable energy systems as a whole does not always dominate in the shaping and implementing of energy policy. It can be stated that there is a lack of initiatives to measure how the policy implemented affects sustainability of the energy sector as a whole. Despite the fact that there is a considerable attempt to assess sustainability of the sector in different perspectives at the scientific level, such evaluations are still very limited in practice. Most often, the level of the target implementation is assessed. The significance of different tools, models, and methods for the energy demand modelling and energy sector planning has increased over the last decades. Scientists are seeking to a model and forecast the most efficient ways toward transition to a sustainable and low-carbon energy system.

The growing public pressure and various political documents and globally implemented strategies (for example, the 2015 Paris Agreement, the United Nations' Sustainable Development Goals [SDGs]) have encouraged many countries to set national renewable electricity, transport, or heating/cooling targets. For example, in 2005 only 43 countries had national renewable energy targets while, after 10 years the number of countries increased to 164 (IRENA, 2019). According to the projections, the consumption of electricity in industry, transportation, and buildings will increase by around 40 percent in 2050 (IRENA, 2019).

DOI: 10.1201/9781003327196-3

The electricity sector has the greatest potential to reduce GHG emissions in many countries (Eskeland et al., 2012). For example, a lot of studies were performed in order to analyze the potential to reduce emissions in the European Union (EU). Corsatea and Giaccaria (2018) performed CO_2 emissions simulation of 13 EU member states, which emit for about 40 percent of CO_2 in the region. According to the results, the technological efficiency increase might help to decline CO_2 emissions for about 5.6 percent. Such decrease would allow to reduce CO_2 emissions in electricity generation sector by 90 percent until 2050 (if compared with 1990) and the EU target regarding GHG emissions would be met. The research by Knopf et al. (2015) focused on the analysis of economic costs and investments in renewable-electricity. The results revealed that the share of renewable energy sources (RES) in the electricity sector should be 49 percent, that the target set by the EU to generate 27 percent of energy from RES by 2030 would be implemented. Also, according to the results, the cost-effective RES-electricity share varies from 43 percent to 56 percent in the EU countries, depending many aspects as economic situation, infrastructure, available energy capacities, public perception, and so forth. The electricity sector accounts for a significant share of final energy consumed and has a huge potential to use more clean energy. Therefore, huge attention should be paid to the assessment of efficiency of the policy implemented and monitoring the progress achieved.

Decarbonization of the heating sector is another important part of the global low-carbon energy transition, and an extremely important step toward achievement commitments of climate change mitigation (Frank et al., 2020). Currently, energy needs for heating and cooling count for about 40–50 percent of global energy. The heating sector is one of the main GHG emissions emitters and also has one of the biggest potentials to use RES. In the EU, heating and cooling energy accounts for half the energy consumed. Heating and preparation of hot water account for 80 percent of total final energy use in the household sector. The industry sector employs 70 percent of consumption for buildings and business process heating, the other 27 percent for technological processes and lighting, 3 percent for cooling. It is important to mention, that three fourths of energy for heating and cooling is generated from fossil fuels in the EU (Eurostat, 2022). Therefore, the EU heating and cooling sector is described as low-efficiency and fossil based. Due to technologies on which this sector relies (Kavvadias et al., 2019), heating and cooling sector decarbonization is recognized as one of the most important objectives to achieve targets of emission reduction and energy efficiency in the EU. Many initiatives are being taken at the EU and national levels for the decarbonization of the heating and cooling sector. Admittedly, this sector has received attention quite late. It is not enough to measure and analyze only the degree of GHG emission reduction or share of RES. Heating related indicators reveal not only climate change but also social and economic issues. For example, in 2020, an average of 8.2 percent of the EU population were unable to heat their homes enough, 6.3 percent of population cannot pay energy bills on time, and 14 percent of the population lived in energy inefficient houses (Eurostat, 2022). Therefore, in order to implement sustainability and climate change goals in the heating sector it is important to consider not only environmental, but also social and economic aspects.

In the world, about 25 percent of all energy generated is consumed in the transport sector. In Europe, the portion of energy to meet transport needs is significantly larger and accounts for about one third. Therefore, in order to implement measures regarding climate change and the decarbonization of the energy sector, it is very important to implement measures in all energy sectors where transport plays a significant role. More than 25 percent of total GHG emissions in the EU comes from the transport sector. Also, the energy demand in transport clearly tends to rise each year. Therefore, the creation of a smart, safe, competitive, accessible, and affordable transport system is one of the main EU energy policy objectives. According to the statistical data (Eurostat, 2022), only seven EU member states have a negative change in total GHG emissions from transport in the years 1990–2020 (Sweden, Italy, Finland, Germany, France, Estonia, and the Netherlands). However, it should be noted, these results are also affected by the COVID-19 lockdowns. Because until 2019, the growth of emissions from transport sector tended to increase each year and only in 2020 has a decrease been observed. The statistical data before the pandemic shows that only three countries had a negative change in the period 1990–2019, these are: Sweden (-15.24%), Finland (-7%) and Estonia (-3.23%). While, the overall GHG emissions in the EU increased by 24 percent in the period 1990–2019. Also, it is necessary to highlight, that sustainable transport development has clear linkages with the well-being of people, such as health, energy poverty, or quality of life.

In order to achieve ambitious energy policy goals and to fight against climate change it is necessary to measure and compare countries' achievements in their progress toward sustainability and low-carbon transition. For that, indicators play an important role, which denotes sustainability of the sector in a holistic way. Sustainability assessment allows not only for ranking countries by the progress achieved, but also to analyze the measures implemented and to follow good examples in other countries. The development of a comprehensive set of indicators for sustainability assessment allows for measuring the effectiveness of political initiatives and support tools and for applying the best practices of the countries to make decisions in more sustainable and effective ways in the future.

3.2 PRACTICAL EXAMPLES OF MULTI-CRITERIA ANALYSIS APPLICATION

This subsection provides practical illustrative examples of multi-criteria analysis application for the assessment and monitoring of policy progress by achieving energy policy goals for the assessment of electricity, heating and transport sector sustainability and ranking of the countries by achievements is made. All created frameworks and assessment procedures were applied for the EU member states.

3.2.1 FRAMEWORK TO MONITOR POLICY PROGRESS BY ACHIEVING ENERGY POLICY GOALS

This subsection presents the framework for the assessment of achievements regarding policy objectives. The framework serves not only as an assessment tool, but also allows for monitoring the progress achieved. The case study provided analyzes the trends of energy development across the eight Baltic Sea Region (BSR) countries. The analysis covers the period of 2008–2015. The calculations and results achieved are analyzed by the application of innovative multi-criteria decision making (MCDM) technique. The EU energy policy priorities govern the selection of indicators for the analysis.

3.2.1.1 Energy Policy Context

Sustainable energy development is a crucial principle in European energy policy. The development of RES gives an opportunity to create a secure, competitive, and sustainable energy sector. Also, it allows for solving the most acute energy problems and challenges facing each EU member state, such as reducing energy dependency, increasing the level of energy security, and achieving climate change and energy goals set by various political documents and strategies (Pacesila et al., 2016).

Reducing energy dependency on imported energy, especially dependency on natural gas and oil from Russia is one of the main priorities of the EU, which was also encouraged by the Russian invasion of Ukraine in the beginning of 2022 (European Commission, 2022a). The EU countries rely on imported energy sources: 57.5 percent of the energy consumed in the region was imported in 2020. Russia is the biggest supplier of natural gas, coal, and oil to the EU. The largest net importers of energy are the EU countries with the biggest populations, except Poland (because of national coal reserves). The level of the EU's energy security is low because of high concentration of import in a few hands. For example, about two thirds of the EU natural gas comes from Russia, Norway, and Algeria. In a similar situation is solid fuel imports, where the main suppliers are Russia, Colombia, and the United States. Dependency on energy imports increased from 40 percent in 1990 to 57.5 percent by 2020 of gross energy consumption (Eurostat, 2022).

Increase energy security in the supply. The portion of imported energy has been rising in the EU over the recent three decades. Now, energy security issues are one of the most significant challenges of EU energy policy. The EU is expected to increase energy security by various measures for energy markets and to prevent and mitigate the possible consequences of potential disruptions. For example, one of the initiatives is the strategy, Energy 2020. This strategy defines energy priorities for a ten-year period and puts forward actions that can be taken to solve climate change and energy policy challenges: create an internal energy market with secure energy supplies and competitive energy prices; increase the level of technological achievements; and effectively interact with supply partners (European Commission, 2010a).

The other initiative is the establishment of an international organization with the Energy Community in 2005. The organization brings together the EU and its neighboring countries with the aim to create an integrated European energy market. The main objective of the organization is to extend the EU internal energy market regulations and principles to countries in the Black Sea region and Southeast Europe. Today, the Energy Community has nine contracting parties. A crucial role in increase of energy security are the diversity of suppliers, energy mix, and energy transport routes. There are many ongoing initiatives to develop electricity corridors and gas pipelines between European countries and their neighbors (Energy Community, 2022).

In order to achieve considerable improvements, the Energy Security Strategy was adopted in 2014. The main goal of the strategy is to ensure energy supply for the European energy consumers. The most crucial issue of energy security is the dependency on a single non-EU supplier. This aspect is especially urgent for gas and oil. Also, it is very important for electricity. For example, the Baltic States are still dependent on external supplier of their electricity network operation and balancing the electricity. The electricity network interconnection between Lithuania and Poland (LitPol Link) is at the top of the agenda for the Baltic States. The Energy Security Strategy proposes actions in several areas: implementation of energy efficiency targets to reach the 2030 climate change and energy policy goals; demand management through innovative and smart technologies (smart meters, billing information); creation of an internal energy market and the development of interconnections to ensure high security supply levels; diversification of energy suppliers, including the increase of produced energy in the EU; strengthening the mechanism of information exchange with the EU government institutions and third countries regarding planned agreements; strengthening the essential energy infrastructure and internal market unity and emergency and mechanisms (European Commission, 2014a). The indicators for monitoring the progress toward a secure European energy market is presented in Table 3.1.

The rate of net import dependency indicates how much a country is dependent on imported energy to meet national energy needs. The statistical data show that all EU countries imported energy in 2020. The import dependency rate and the change in the last 14 years varies across countries and depends on many factors, especially national resources, policy implemented and the output of the country.

The aggregate supplier concentration index indicates the extent of energy supplied from countries outside the European Economic Area (EEA). Small values of concentration indicate high diversification of national energy sources and can be assumed as a condition for lower risk to face an energy supply crisis. While higher values of concentration show the low level of energy sources diversification and, accordingly, constitute an important factor increasing risk to face an energy supply problems. For the whole EU, the supplier concentration index is quite low. It can be said, that imported energy is diversified in the region. But looking at individual member states level, the index varies from less than 10 to almost 80. Countries that generate a lot of local energy, like Denmark, and countries that mostly import energy from EEA countries, like Luxembourg, or countries with a relatively low consumption of natural gas, coal, and oil, and countries having many various suppliers, like France, have a low index. In a few EU member states the index is very high (more than 50) and these countries are mostly dependent on suppliers from Russia, like Bulgaria, Estonia, and Finland. The level of the aggregate supplier concentration index in Cyprus was almost zero in 2020 because the energy for Cyprus was supplied from the EU internal market.

The N-1 rule for gas shows the national adequacy and security of gas infrastructure. It indicates the ability to satisfy the gas demand in case of a disruption of the single largest gas infrastructure item during times of extremely high energy demand (e.g., cold winter days). The indicator is expressed as the percentage of total demand that can be fulfilled with the remaining infrastructure. As mentioned before, natural gas raises one of the biggest concerns about the security of supply in the whole

TABLE 3.1

The EU Energy Security and Import Dependency Indicators

	Net import dependency rate		Supplier concentration index		N-1 rule for gas	
Country	Net imports, % of gross inland consumption and international bunkers, 2019	Absolute change 2005 – 2019, %	%, where 100 means maximum concentration, 2018	Absolute change 2005 – 2018, %	% of total demand that can be satisfied if the largest item of gas supply infrastructure is disrupted, 2019	Absolute change 2009 – 2019, %
EU	60.62	2.8	12.84	2.61	NA	NA
AT	71.73	–0.03	36.07	13.3	130	–15
BE	76.68	–3.29	15.1	–2.73	273	76
BG	38.1	–9.24	57.11	2.46	36.2	7.2
HR	56.22	3.66	9.37	–25.62	100.6	NA
CY	92.81	–7.9	0.27	–1.16	NA	NA
CZ	40.59	12.75	31.63	5.88	299.7	152.7
DK	38.76	89.37	8.87	4.65	100	27
EE	4.83	–23.21	76.05	26.68	105	–39
FI	42.09	–12.37	67.05	1.55	125.4	125.4
FR	47.59	–4.1	9.89	2.59	131	28
DE	67.61	6.87	24.67	12.43	227	–18
GR	68.86	0.66	38.02	–1.48	112.4	–122.6
HU	69.7	7.45	69.75	23.75	143	62
IE	68.39	–21.26	14.82	–9.65	108	90
IT	77.48	–5.86	19.19	3.55	88	–36
LV	43.96	–19.88	42.32	–6.66	248.59	85.59
LT	75.22	19.89	47.75	–48.75	153.4	96.4
LU	95.13	–2.25	5.29	–22.28	77.6	–29.4
MT	97.17	–2.8	10.73	10.73	NA	NA
NL	64.72	26.94	15.44	7.35	206	41
PL	46.82	29.07	25.19	–0.8	118.2	–15.8
PT	73.85	–14.71	24.65	3.34	114	9
RO	30.37	2.9	25.77	7.74	100.7	5.7
SK	69.76	3.77	68.17	–5.48	323.5	205.5
SI	52.14	1.31	1.63	–24.28	66.3	–8.7
ES	74.96	–6.57	17.25	3.62	127	3
SE	30.24	–7.83	14.55	–0.29	2.5	–7.5

Source: Created by authors based on data from European Commission (2022b).

EU. Gas supply security requires that all countries ensure the fulfilment of the total consumers' energy needs in the case of the biggest gas infrastructure disruption. This condition is met if indicators reach at least 100 percentages. In 2020 only a few EU countries showed values of indicators less than 100 percent: Bulgaria, Italy, Luxembourg, Slovenia, and Sweden.

Meet climate change and energy targets. The main contributor to GHG emissions in Europe is the energy sector, where carbon dioxide is released into the atmosphere because of fossil fuel combustion. The development of RES and improvements in energy efficiency are the main ways to reduce emissions in the energy sector. The EU sought to achieve at least 20 percent of the final energy consumption from RES by 2020, and aims for at least 27 percent by 2030. The target for 2020 is already successfully implemented. Also, all member states achieved their national RES targets, which were set by the Renewable Energy Directive (2009/28/EC) (European Parliament and Council of the European Union, 2009), taking into account the different situations of the countries and possibility to implement targets. The EU countries were free to choose how they support the RES development. The most common support mechanism for RES in the electricity sector were feed-in tariffs, competitive auctions, or feed-in premiums in the beginning of target implementation. Table 3.2 provides indicators allowing for monitoring progress of energy decarbonization in the EU.

The reduction of GHG emissions is another fundamental objective of the EU energy policy. The EU has set ambitious targets for the decarbonization of its economy with the aim to reduce GHG emissions at least 80 percent by 2050. Various scientific studies show that achievement of such reduction requires huge structural changes of the energy system (Spencer et al, 2015; Bataille et al. 2016; Spencer et al., 2017; OECD/IEA, IRENA, 2017). The mid-term target to reduce emissions by at least 20 percent by 2020 is already reached. The most important policy tools to achieve the GHG target are the EU Emissions Trading System (ETS) and the Effort Sharing Decision (ESD). In 2016 the Clean Energy for All Europeans (European Commission, 2016a) was presented by the European Commission. The document proposed a package of measures with the aim to create a stable legislative framework with fully updated energy scenarios for deep emission reductions, and legislative action to encourage clean energy transition until 2030. The EU cut GHG emissions about 24 percentage points compared with the 1990 level, by 2020 (Eurostat, 2022). However, a lot of work and effort remains to be done in the future.

Energy efficiency targets also have been set and various measures improving energy savings in the EU have been adopted, such as: no less than 3 percent of governmental buildings should be renovated in energy efficient ways each year; mandatory buildings efficiency certificates; standards for energy efficiency, and labelling for products; preparation of national action plans for energy efficiency improvements for each 3 years; development of smart meters; energy consumption audit for large companies each 4 years; ensuring energy consumers rights to have free access to information. In 2016 the European Commission proposed an update to the Energy Efficiency Directive (2016/0376) (European Commission, 2016b), where a 30 percent energy savings target for 2030 was presented, and measures to ensure and to monitor progress were provided.

3.2.1.2 Indicators for Comparative Assessment of EU Energy Policy Priorities Implementation

At both the national and international levels, a large number of studies have been performed to assess sustainable energy development in the past three decades (Hirschberg et al., 2007; Brown and Sovacool, 2007; OECD, 2013; Iddrisu and

TABLE 3.2
Indicators to Monitor Progress toward the Decarbonization of the EU Energy Sector, 2019

Country	Primary energy consumption		Final energy consumption		GHG emissions reduction		Renewable energy share	
	2019, Mtoe	Average annual change 2005-2019, %	2019, Mtoe	Average annual change 2005-2019, %	2019, %	Gap between GHG projections and 2020 target in Effort Sharing sectors *, 2020, %	2019, %	2020 target implementation, 2020, %
EU	1351.93	-0.7	983.59	-0.37	76.34	5.55	19.9	99.5
AT	32.2	-0.06	28.28	0.15	101.23	3.44	36.55	107.49
BE	49.11	-0.24	35.76	-0.07	79.52	-5.01	13.00	100.00
BG	18.22	-0.26	9.84	-0.09	54.98	-19.23	23.32	145.74
HR	8.21	-0.72	6.91	-0.29	75.25	2.92	31.02	155.12
CY	2.54	0.32	1.89	0.3	136.81	-6.5	16.88	129.84
CZ	40.12	-0.37	25.24	-0.23	65.28	1.77	17.30	133.10
DK	16.83	-0.95	14.34	-0.52	61.55	-5.02	31.65	105.49
EE	4.71	0.06	2.9	0.15	37.19	1.43	30.07	120.28
FI	32.06	-0.22	25.32	0.11	73.06	0.75	43.80	115.27
FR	235.26	-0.72	145.45	-0.65	79.12	5.38	19.11	83.08
DE	282.71	-0.85	214.54	-0.09	63.78	-21.03	19.31	107.29
GR	24.26	-1.48	16.19	-1.73	80.51	-18.43	31.75	120.83
HU	24.57	-0.45	18.61	0	66.71	17.99	13.85	106.54
IE	14.66	-0.06	12.36	-0.06	103.88	-6.84	16.16	101.00
IT	145.89	-1.48	115.5	-1.2	79.87	-11.67	20.36	119.76
LV	4.56	0.16	4.08	0.19	43.45	-6.81	42.13	105.33
LT	6.28	-1.55	5.56	1.33	41.76	2.73	36.77	116.40
LU	4.5	-0.38	4.39	-0.1	81.67	26.54	11.70	106.35
MT	0.87	-0.07	0.7	3.08	76.68	-10.36	10.71	107.14

NL	63.46	-0.66	49.89	-0.5	81.28	-3.97	14	99.99
PL	98.13	0.85	70.97	1.45	82.12	-24.72	16.10	107.35
PT	22.08	-0.81	17.13	-0.71	105.76	-17.6	33.98	109.62
RO	31.97	-0.75	23.88	-0.15	45.09	-19.74	24.48	101.99
SK	15.98	-0.52	11.17	-0.14	56.64	-11.52	17.35	123.89
SI	6.52	-0.66	4.85	-0.31	91.51	-4.13	25	100.00
ES	120.76	-0.79	86.3	-0.87	106.57	-15.28	21.22	106.10
SE	45.78	-0.38	31.57	-0.33	68.59	N/A	60.12	122.70

Source: Created by authors based on data from European Commission (2022b); Eurostat (2022).

Note: the abbreviations of the EU countries can be found in Appendix 3.1.

* of 2005 base year emissions

Bhattacharyya, 2015; World Economic Forum, 2015; RSC project, 2016; OECD, 2003; OECD/NEA, 2002; Burgherr et al., 2005; Zelazna and Golebiowska, 2015; Streimikiene and Siksnelyte, 2016; Sartori et al., 2017, etc.). Different indices were created, and many different indicators were used to reflect specific characteristics among countries. It allows for comparing countries and gives a comprehensive picture of the energy system. The changes of indicators over time are characteristics of progress made and indicate the efficiency of policy implemented. Various indicators help decision makers to measure the effectiveness of the support mechanisms, to evaluate the progress and help to guide decisions on investments or air pollution control, and so forth.

The International Atomic Energy Agency, the International Energy Agency, United Nations Department of Economic and Social Affairs, Eurostat and the European Environment Agency have created a set of indicators reflecting sustainable energy development – Energy Indicators for Sustainable Development (EISD). The set consists of 30 indicators, organized into three groups, reflecting three sustainability dimensions: economic, social, and environmental. These dimensions are classified into seven themes and nineteen sub-themes. Also, it should be noted, that some indicators can be classified in more than one dimension and theme or sub-theme. This shows the undeniable interlinkages among these indicators (IAEA, 2005). The structure of EISD is presented in Table 3.3.

TABLE 3.3
The Structure of EISD Indicators

Dimension	Theme	Sub-theme
Economic	Use and Production Patterns	• Overall energy use
		• Overall energy productivity
		• Energy production
		• End-use
		• The efficiency of energy supply
		• Energy prices
		• Energy diversification
	Security	• Imported energy
		• Strategic fuel stock
Environmental	Atmosphere	• Climate change
		• Quality of air
	Water	• Quality of water
	Land	• Quality of soil
		• Impact to forest
		• Solid waste management and generation
Social	Equity	• Energy affordability
		• Energy accessibility
		• Disparities
	Health	• Safety

Source: Created by authors.

In order to compare countries' achievements in sustainable energy development, the indicators were selected based on the reliability and availability of data and with the purpose to reflect the EU sustainable energy development objectives. The indicator set and their weights for the assessment are provided in Table 3.4.

TABLE 3.4
Set of Indicators to Monitor Policy Progress by Achieving Policy Goals

Impact area	Label	Indicator	Measurement	Target value	Weight
Economic dimension (1/3)					
Use and Production Patterns	EC1	Overall energy use	Energy per Capita, kg	min	1/21
	EC2	Energy productivity	Total primary energy use per unit of GDP	min	1/21
	EC3	Energy intensity	Primary energy intensity – toe/M€'	min	1/21
	EC4	Supply efficiency	Distribution losses, % of energy generated	min	1/21
Energy security	EC5	Energy independence	Import independency, %	max	1/21
	EC6	Supplier concentration in electricity sector	Cumulative market share in generation, Main entities, %	min	1/21
	EC7	Supplier concentration in gas sector	Cumulative market share, Main entities, %	min	1/21
Environmental dimension (1/3)					
Atmosphere	EN1	Reduce GHG emissions	Target implementation, %	max	1/15
	EN2	Increase in the share of RES in final energy consumption	Target implementation, %	max	1/15
	EN3	Increase in primary energy efficiency	Target implementation, %	max	1/15
	EN4	Increase in final energy efficiency	Target implementation, %	max	1/15
Land	EN5	Energy consumption (waste (non-RES))	% of gross inland consumption	max	1/15
Social dimension (1/3)					
Equity	SO1	Affordability of electricity in households' sector	Price, EUR	min	1/15
	SO2	Affordability of gas in households' sector	Price, EUR	min	1/15

(*continued*)

TABLE 3.4 (Continued)
Set of Indicators to Monitor Policy Progress by Achieving Policy Goals

Impact area	Label	Indicator	Measurement	Target value	Weight
	SO3	Opportunity to choose gas supplier	Electricity retailers to final consumers –Nr, the base year 2005	max	1/15
	SO4	Opportunity to choose electricity supplier	Gas retailers to final Consumers –Nr, the base year 2005	max	1/15
Health	SO5	CO2 emissions	CO2 per capita, kg	min	1/15

Source: Created by authors.

In this assessment, all dimensions of sustainability are equal. The weights of indicators in each dimension also are equal. Despite that, it is possible to change the weights for future research easily in order to monitor the achievements made or to measure the sensitivity of each dimension. The level of the target implementation has been selected to indicate countries' different situations and ability to achieve the sustainable energy objectives. Four of five environmental indicators define implementation of EU energy policy priorities (increase the share of RES, reduce GHG emissions, increase energy efficiency). Seven economic indicators were selected to monitor the progress of energy security and patterns of energy use in the region. Five social indicators were chosen to reflect the most important social issues of energy policy. The developed assessment framework was applied for the measurement of achievements of Baltic Sea Region (BSR) countries in the period of 2008–2015 (Siksnelyte et al., 2019). Table 3.5 provides the summary of selected indicators for 2008.

Table 3.6 presents countries achievements in 2015.

3.2.1.3 MCDM Technique

The method selected for the assessment is governed by the synthesis of the neutrosophic sets (Smarandache, 1999) and traditional Full Multiplicative Form of Multi-Objective Optimization by Ratio analysis (MULTIMOORA) technique proposed by Brauers and Zavadskas (2010). The aggregated decision making matrix is constructed as is usually done in the MCDM framework. The x_{ij} elements correspond to i^{th} criteria of jth alternative. The matrix is constructed as follows:

$$X = \begin{bmatrix} x_{11} & \cdots & x_{1m} \\ x_{1n} & \cdots & x_{nm} \end{bmatrix} \qquad (3.1)$$

TABLE 3.5
The Summary of Indicators of BSR Countries, 2008

	Denmark	Estonia	Finland	Germany	Latvia	Lithuania	Poland	Sweden
EC1	3593.6	4441.0	6787.7	4101.2	2141.6	2888.3	2568.3	5369.3
EC2	0.080	0.347	0.179	0.123	0.188	0.249	0.253	0.133
EC3	77.4	339.8	175.3	119.8	212.0	251.8	273.1	127.9
EC4	5.12	5.84	3.22	1.91	4.00	4.59	2.95	3.62
EC5	120.47	75.29	45.90	39.06	41.19	42.22	69.75	62.94
EC6	78.0	96.5	63.0	72.0	93.0	87.6	45.9	80.0
EC7	100.0	100.0	100.0	89.0	100.0	99.5	96.2	100.0
EN1	78.14	92.69	85.34	86.49	109.41	114.11	107.43	88.83
EN2	62.0	75.6	82.4	47.8	74.5	77.4	51.3	92.5
EN3	88.51	112.31	103.34	86.26	114.82	73.85	103.73	91.71
EN4	92.36	89.29	103.75	88.01	106.67	81.40	112.85	93.07
EN5	2.266	0.000	0.322	0.985	0.107	0.000	0.246	0.880
SO1	0.264	0.081	0.122	0.215	0.084	0.086	0.126	0.170
SO2	0.096	0.034	0.033	0.064	0.031	0.033	0.042	0.093
SO3	3.2	1.3	1.0	1.0	1.0	1.2	1.0	0.9
SO4	0.514	0.925	1.000	1.000	1.000	1.143	0.517	0.926
SO5	9946.3	13415.6	11408.8	10 687.1	3871.6	4765.4	8630.6	5801.7

Source: Created by authors.

TABLE 3.6
The Summary of Indicators of BSR Countries, 2015

	Denmark	Estonia	Finland	Germany	Latvia	Lithuania	Poland	Sweden
EC1	2962.3	4757.1	6059.3	3869.6	2205.3	2366.4	2511.1	4665.2
EC2	0.061	0.304	0.152	0.097	0.175	0.155	0.209	0.098
EC3	64.1	352.4	170.8	105.0	201.4	172.3	214.4	107
EC4	5.514	4.702	2.717	1.801	3.274	3.861	2.469	2.374
EC5	86.88	92.62	53.18	38.09	48.85	21.56	70.70	69.89
EC6	44.0	79.8	62.8	76.0	57.4	63.2	25.5	73.4
EC7	100.0	100.0	100.0	80.3	100.0	93.1	82.5	100.0
EN1	98.57	98.01	95.16	91.93	109.81	113.06	108.97	106.04
EN2	102.67	114.40	103.42	81.11	94.00	112.17	78.67	110.00
EN3	105.17	104.62	110.86	94.11	120.37	110.77	106.64	99.31
EN4	103.47	100.00	109.36	90.84	115.56	86.05	112.99	95.05
EN5	2.547	1.071	0.721	1.353	1.256	0.333	0.548	1.370
SO1	0.307	0.130	0.155	0.295	0.164	0.126	0.144	0.185
SO2	0.080	0.046	0.054	0.068	0.05	0.042	0.05	0.113
SO3	3.40	0.91	0.77	1.35	4.00	0.80	1.54	1.00
SO4	0.70	1.15	1.00	1.32	1.00	2.43	0.51	0.97
SO5	6746.8	12137.4	8479.3	10054.1	3817.9	4582.6	8222.8	4669.0

Source: Created by authors.

In the first stage, for the normalization of the aggregated decision matrix the vector normalization approach is applied:

$$X^* = \frac{x_{ij}}{\sqrt{\sum\limits_{i=1}^{m} x_{ij}^2}} \tag{3.2}$$

After this stage, the aggregated decision matrix is presented in the neutrosophic form. The first objective of neutrosophic aggregative MULTIMOORA technique will have the following expression:

$$Q_j = \sum_{i=1}^{g} w_i \left(x_n^*\right)_{ij} + \left(\sum_{i=g+1}^{n} w_i \left(x_n^*\right)_{ij}\right)^c \tag{3.3}$$

where g elements express the members of the criteria maximized, and $n\text{-}g$ components correspond to the criteria minimized. Normally, single-valued neutrosophic members have the structure as (Zavadskas et al., 2017):

$$\left(x_n^*\right)_1 = \left(t_{n1}, i_{n1}, f_{n1}\right) \tag{3.4}$$

When the second objective is constructed, the deviation from the reference point applying Min-Max Norm of Tchebycheff is considered:

$$\min_j \left(\max_i \left| D\left(r_i - w_i \left(x_n^*\right)_{ij}\right) \right| \right) \tag{3.5}$$

The reference point is expressed as follows:

$$r_i = (1.0;\ 0.0;\ 0.0) \tag{3.6}$$

For the case of the maximized and minimized criteria:

$$r_i = (0.0;\ 1.0;\ 1.0) \tag{3.7}$$

The score function is applied to the relating the neutrosophic members as follows:

$$S\left(\left(x_n^*\right)_1\right) = \frac{3 + t_{n1} - 2i_{n1} - f_{n1}}{4} \tag{3.8}$$

The distance between two single-valued neutrosophic members is calculated by applying the following function:

$$D\left(\left(x_n^*\right)_1, \left(x_n^*\right)_2\right) = \sqrt{\frac{1}{3}\left(\left(t_{n1} - t_{n2}\right)^2 + \left(i_{n1} - i_{n2}\right)^2 + \left(f_{n1} - f_{n2}\right)^2\right)} \tag{3.9}$$

At the last stage, the third objective is constructed by a Full Multiplicities form which includes maximized and minimized criteria expressed by the purely multiplicative utility function. Then, the overall utility for each considered alternative can be expressed by the equation presented below:

$$U_j = \frac{S(A_j)}{S(B_j)}$$
(3.10)

Here A_j and B_j components are calculated as:

$$A_j = \prod_{i=1}^{g} w_i \left(x_n^*\right)_{ij}, \quad B_j = \prod_{j=g+1}^{n} w_i \left(x_n^*\right)_{ij}$$
(3.11)

Where A_j presents the product of maximized criteria of jth alternative. Accordingly, B_j symbolizes product of minimized criteria of the same alternative. And finally, all three objectives are summarized by the theory of dominance (Brauers and Zavadskas, 2011):

3.2.1.4 Multi-Criteria Assessment Results

The results of the neutrosophic ratio system for the BSR countries are presented in Tables 3.7 and 3.11. The results of the neutrosophic reference point for the countries below are provided in Tables 3.8 and 3.12. The neutrosophic full multiplicative form for the BRC countries is shown in Tables 3.9 and 3.13. The dominance theory was applied to receive the final evaluation by all neutrosophic MULTIMOORA technique objectives, and the results can be found in Tables 3.10 and 3.14.

The leading countries in BSR by the achievements in sustainable energy development are Denmark and Latvia. In 2008, Denmark was ranked first, Latvia was in the second position (Table 3.15). In 2015 these two countries changed their positions.

TABLE 3.7
The Neutrosophic Ratio System Objective, 2008

Country	Q_i			$S(Q_i)$	Rank
Denmark	(0.8497	0.1451	0.1493)	0.8526	4
Estonia	(0.8092	0.1988	0.2056)	0.8015	8
Finland	(0.8378	0.1659	0.1798)	0.8315	7
Germany	(0.8571	0.1452	0.1676)	0.8498	5
Latvia	(0.8980	0.1043	0.1264)	0.8908	1
Lithuania	(0.8801	0.1214	0.1462)	0.8728	2
Poland	(0.8775	0.1234	0.1475)	0.8708	3
Sweden	(0.8422	0.1611	0.1781)	0.8355	6

Source: Created by authors.

TABLE 3.8
The Neutrosophic Reference Point Objective, 2008

Country	$\max\left\lvert D\left(r_i - w_i\left(x_n^*\right)_{ij}\right)\right\rvert$	Rank
Denmark	0.9865	1-2
Estonia	0.9997	7-8
Finland	0.9917	4
Germany	0.9901	3
Latvia	0.9973	6
Lithuania	0.9997	7-8
Poland	0.9935	5
Sweden	0.9865	1-2

Source: Created by authors.

TABLE 3.9
The Neutrosophic Full Multiplicative Form Objective, 2008

Country	$S(A_i)$	$S(B_i)$	U_i	Rank
Denmark	0.3156×10^{-16}	0.0380×10^{-15}	0.82990	1
Estonia	0.0002×10^{-16}	0.3299×10^{-15}	0.00006	8
Finland	0.0017×10^{-16}	0.0620×10^{-15}	0.00275	6
Germany	0.0014×10^{-16}	0.0153×10^{-15}	0.00945	5
Latvia	0.0006×10^{-16}	0.0017×10^{-15}	0.03509	2
Lithuania	0.00004×10^{-16}	0.0090×10^{-15}	0.00044	7
Poland	0.0014×10^{-16}	0.0082×10^{-15}	0.01671	3
Sweden	0.0078×10^{-16}	0.0574×10^{-15}	0.01356	4

Source: Created by authors.

The biggest drop in the positions was determined in Poland and Germany during the period considered. Germany, which was ranked fifth among the BSR countries in 2008, took last place in 2015. Poland dropped from third to sixth in the ranking, while Estonia and Lithuania made the biggest progress during the period under analysis.

Each country under analysis is discussed separately to illustrate the progress made in implementing goals of sustainable energy development.

Denmark and Latvia were the leading countries in sustainable energy development and implementation of EU energy policy goals. Despite that Denmark systematically sought to implement the EU's energy policy objectives, some indicators dropped during the period under analysis, and Latvia rose to first place in 2015. Since 2013, Denmark has been the only country in the region that had negative import dependency. Also, the country is distinguished by the highest energy prices in the BSR.

TABLE 3.10
The Rankings of the BSR Countries by Neutrosophic MULTIMOORA Method, 2008

Country	The neutrosophic ratio system	The neutrosophic reference point	The neutrosophic full multiplicative form	Final rank
Denmark	4	1-2	1	1
Latvia	1	6	2	2
Poland	3	5	3	3
Sweden	6	1-2	4	4
Germany	5	3	5	5
Finland	7	4	6	6
Lithuania	2	7-8	7	7
Estonia	8	7-8	7	8

Source: Created by authors.

TABLE 3.11
The Neutrosophic Ratio System Objective, 2015

Country	Q_i			$S(Q_i)$	Rank
Denmark	(0.8810	0.1163	0.1293)	0.8798	3
Estonia	(0.8044	0.2046	0.2059)	0.7973	7
Finland	(0.8377	0.1647	0.1811)	0.8318	6
Germany	(0.7814	0.2252	0.2388)	0.7730	8
Latvia	(0.9062	0.0943	0.1154)	0.9006	2
Lithuania	(0.9101	0.0886	0.1106)	0.9056	1
Poland	(0.8750	0.1255	0.1505)	0.8684	4
Sweden	(0.8416	0.1618	0.1760)	0.8355	5

Source: Created by authors.

The ESD sets national annual binding targets for GHG emissions not covered under the ETS. The national targets were determined taking into account the economies of the member states and their individual growth potential. The highest emissions target (-20%) among selected countries was set for Denmark. Danish energy policy is exceptionally related with the EU policy; for example, in 2017 the Danish government announced that Denmark will seek to produce 50 percent of its energy from RES by 2030; the country seeks to reduce GHG emissions by 80–95 percent until 2050. Denmark has already achieved its target for RES in final energy consumption (30%) set by the strategy, Europe 2020, in 2015. The most significant progress during the period considered was found in electricity (biomass and wind energy) and heating (biomass). However, fossil fuels still dominated the energy mix. The country

TABLE 3.12
The Neutrosophic Reference Point Objective, 2015

Country		$\max\left|D\left(r_i - w_i\left(x_n^*\right)_{ij}\right)\right|$	Rank
Denmark	**DK**	0.9868	2
Estonia	**EE**	0.9896	4
Finland	**FI**	0.9911	6
Germany	**DE**	0.9914	7
Latvia	**LV**	0.9825	1
Lithuania	**LT**	0.9938	8
Poland	**PL**	0.9904	5
Sweden	**SE**	0.9886	3

Source: Created by authors.

TABLE 3.13
The Neutrosophic Full Multiplicative Form Objective, 2015

Country	$S(A_i)$	$S(B_i)$	U_i	Rank
Denmark	0.1354×10^{-16}	0.0025×10^{-14}	0.54280	2
Estonia	0.0100×10^{-16}	0.0571×10^{-14}	0.00175	4
Finland	0.0012×10^{-16}	0.0573×10^{-14}	0.00020	7
Germany	0.0011×10^{-16}	0.2665×10^{-14}	0.00004	8
Latvia	0.0339×10^{-16}	0.0004×10^{-14}	0.82550	1
Lithuania	0.0043×10^{-16}	0.0004×10^{-14}	0.10767	3
Poland	0.0006×10^{-16}	0.0074×10^{-14}	0.00079	6
Sweden	0.0021×10^{-16}	0.0223×10^{-14}	0.00093	5

Source: Created by authors.

also achieved both energy efficiency objectives set by the strategy Europe 2020. It is necessary to highlight that the Danish government has set higher national targets for energy efficiency than set by the rest of the EU. Danish dependence on imported energy sources was one of the smallest among the EU member states and reached only 13.1 percent.

Latvia took first position in the ranking in 2015 and left Denmark in second place. The main reasons Latvia was the first according to the indicators selected are as follows: low level of energy consumption; EU targets for climate change and energy were overachieved (except RES); level of GHG emissions are the lowest among countries under analysis. Despite Latvia having a large portion of RES in the final energy mix, the country was dependent on imported energy, especially natural gas from Russia. The opening of a liquefied natural gas (LNG) terminal in Lithuania

TABLE 3.14
The Rankings of the BSR Countries by Neutrosophic MULTIMOORA Method, 2015

	The neutrosophic ratio system	The neutrosophic reference point	The neutrosophic full multiplicative form	Final rank
Latvia	2	1	1	1
Denmark	3	2	2	2
Lithuania	1	8	3	3
Estonia	7	4	4	4
Sweden	5	3	5	5
Poland	4	5	6	6
Finland	6	6	7	7
Germany	8	7	8	8

Source: Created by authors.

TABLE 3.15
The Achievements of the BSR Countries during the Period under Analysis

Country	Final rank, 2008	Final rank, 2015	Change in ranking
Denmark	1	2	−1
Latvia	2	1	+1
Poland	3	6	−3
Sweden	4	5	−1
Germany	5	8	−3
Finland	6	7	−1
Lithuania	7	3	+4
Estonia	8	4	+4

Source: Created by authors.

began to be an alternative in 2015. Also, it is expected to reduce dependence from one supplier with the Gas Interconnection Poland–Lithuania (GIPL) project, which opened on 1 May 2022. The project should boost the energy security level in the region by connecting the Finnish and Baltic region with the Polish energy markets. Now, the interconnector already allows Lithuanian LNG to flow to Poland, which is very important in the current geopolitical context. Latvia had the highest electricity concentration among the BSR countries and was third in the whole EU (after Cyprus and Malta). Despite the attempts to encourage the interconnection between the neighboring countries, there are several weak points, such as the Estonian and Latvian transmission line and the Latvian internal electricity networks. The security of energy supply and the competitiveness of energy market in the country depends on the development of internal energy infrastructure and electricity and gas links. Various projects to connect Baltic states with the EU single energy market were initiated. For example, the modernization of the Inčukalns underground gas storage facility and the electricity networks synchronization with the European networks. The RES target for Latvia set by the strategy Europe 2020 is the highest among Baltic states (40%). In 2015, the target (38%) was almost reached. Regarding energy efficiency targets, in 2015 both are less than maximum level set. The GHG emissions target set for Latvia is higher by 17 percent compared to the 1990 level. In 2015, the level of emissions was below the permissible limit.

Lithuania rose four places in the period under analysis and took third place in the final ranking in 2015. The country can be characterized by low energy prices, low energy consumption levels, and a high level of energy dependency during the period concerned. However, in the last few years, completed energy infrastructure projects in Lithuania and cooperation with neighboring countries have increased the level of energy security in the country. For example, the opening of the LNG terminal in Klaipeda (Lithuania) allows access to the independent source of natural gas, the gas interconnection Lithuania–Poland (GIPL) which integrates the gas markets of the Baltic countries and Finland, the electricity interconnection with Finland via Estlink1 and Estlink2, with Poland via LitPol Link and Sweden via NordBalt – these developments increased the capacity and security of the supply in recent years. Now one of the most important issues is to encourage the synchronization of Baltic states electricity networks with the continental Europe electricity network and to disconnect from post-Soviet Belarus, Russia, Estonia, Latvia, Lithuania (BRELL) electricity ring. The works of synchronization are following the Baltic energy market interconnection plan (BEMIP) and it is expected to end all works by 2025. The Lithuanian energy sector is heavily dependent on imported energy from non-EU suppliers, and therefore it is very important to strengthen energy security in the country and the whole region, especially by seeking new opportunities for energy supply from reliable suppliers. Development of new renewable infrastructure and improvements in energy efficiency are also undoubtedly measures to improve the situation. However, the implementation of these measures requires quite a lot of time. In 2015, Lithuania already overachieved the national target for RES set by the strategy Europe 2020. The heating sector with more than 46 percent of RES had the most significant input, where 44 percent of energy was generated from biomass and waste incineration. Primary energy consumption was less than the set maximum, but the target for final energy consumption has not yet been achieved. For Lithuania as for Latvia, the target for GFHG emissions was set higher (+15%) than the level of 1990. In 2015, the GHG level was lower than the maximum set by the EU.

The Swedish energy system can be characterized as having a low dependency on fossil fuels and based on RES in all sectors, except transport, which is still based on oil. The share of RES in final energy consumption is one of the highest in the EU, and the Europe 2020 target for RES is overachieved. As with Denmark, Sweden also has set high national energy efficiency targets. In the electricity sector, nuclear and energy from RES accounted for 99 percent of gross energy generated in 2015. The Swedish electricity system can be characterized as highly interconnected with other

Nordic countries, as well as Lithuania (the NordBalt project). The electricity sector is successfully liberalized and has high competition in the retail market. The issues of gas security are not a big problem in the country, because gas accounts for only a few percentages in final energy mix. Regarding the target for GHG emissions (17%), the country has already achieved it and exceeded it by 2.18 Mtoe in 2015.

Another country that made the most significant progress during the period under analysis was Estonia. This increase was due to the policy measures applied for strengthening national energy independence, increasing competition in the energy market, and achieving regional energy policy targets. In 2015, the country had already reached its national targets set by the strategy Europe 2020 for RES and energy efficiency. It is worth to highlight that Estonia is the country most independent from imported energy, not only in the BSR but among all EU member states. For example, in 2015, Estonia's imported energy accounted for only 7.4 percent, while the EU average was 54 percent. In 2019 the indicator was even better, and imported energy was only 4.83 percent of gross inland consumption, while the EU average was 60.62 percent.

Poland's economy is one of the most energy-intensive among all EU countries. The growth of the Polish economy led not reduced national energy consumption in almost all economical sectors since 2005. However, the energy system in Poland is based on fossil fuels (about 80% in 2015) and shows a low level of energy efficiency. For example, about 60 percent of the Polish energy infrastructure is older than thirty years; it requires enormous investments for modernization, enlargement of capacity, and improvement of supply efficiency. Although the richness of national energy sources ensures the security of the energy supply, the old and inefficient infrastructure and huge amount of fossil fuels usage challenges implementation of national energy and climate change objectives and economic decarbonization in the whole region. Although national GHG emissions and energy efficiency targets set by the EU did not exceed during the period under assessment, the economy of Poland is growing each year, resulting in an increase in energy consumption and GHG emissions. The GHG emissions target for Poland was set 14 percent higher than the 1990 level. Now, the development of energy infrastructure and new links are the main tasks for the creation of a secure and efficient Polish energy sector. The national RES target for Poland is the lowest among BSR countries and one of the lowest in the whole EU (15%). In 2015, the country implemented about 80 percent of the target. Although, the target was achieved in 2020, the country had difficulties in development of new renewable infrastructure due to financing, administrative difficulties and the legal basis.

Finland was at the bottom of the ranking regarding the implementation of EU energy policy priorities during the period under assessment. In 2008–2015, Finland notably has had high energy consumption, no competition in the gas market, and quite high level of GHG emissions. Regarding the national RES and energy efficiency target, they had reached it already in 2015. The development of new renewable infrastructure and energy efficiency measures helped to reduce the energy dependency rate, which was 46.8 percent in 2015. Finland consistently forms its energy policy following EU strategies and the objectives of the energy and climate change policy. In 2016, the National Energy and Climate Strategy by 2030 was approved and measures to achieve energy and climate change objectives were defined. According to the strategy, it is expected that RES accounts for 50 percent of the final energy consumption in 2030 (of which 30% for transport). According to recent statistics, the RES accounted for 43 percent in 2020. So the statistics are promising and the country is going systematically toward its set goal. However, it could be a big challenge for Finland to encourage the use of advanced biofuels and promote electrification of transport. There was no competition in the gas market, and the market was under strong regulation during the period under assessment. But, later, the situation was solved by the construction of the Baltic connector pipeline (gas pipeline between Finland and Estonia), which has been in commercial use since the beginning of 2020, and LNG terminals: port of Pori (2016), Tornio Manga (2019), the port of Hamina (2022).

Germany slipped down three positions during the period analyzed and was the last among countries under analysis regarding the implementation of EU climate change and energy policy priorities. The country was strongly dependent on imported energy, notably by low competition in the energy market, EU energy policy priorities were implemented very slowly, and a high level of GHG emissions was observed. The energy sector in Germany was based on fossil fuels during the period under assessment. In 2015 more than 60 percent of its energy came from imported sources, and the level of energy dependency of the country is one of the largest in the BSR. Energy security issues are very actual in the country. Now, in the face of the global energy crisis and economic uncertainty, the consequences of slow development of reliable infrastructure and the search for a reliable energy supplier can be seen. The economy of the country is one of the most energy-intensive in the whole EU, therefore to reduce energy consumption is a very difficult issue. During the period under assessment German energy consumption did not decrease in almost all economic sectors.

The presented framework is developed for the analysis of the EU energy policy achievements. But the presented framework can be easily modified to assess energy policy achievements of other regions or countries. The whole world faces similar climate change and energy policy problems. Therefore, the policy goals of countries seeking sustainability in the energy sector are quite similar, and the proposed system can be applied.

3.2.2 Framework to Monitor Electricity Sector Sustainability

This subsection presents the framework for the assessment and monitoring of electricity sector sustainability. The developed framework was applied in practice for the assessment and ranking of the EU member states. An illustrative example for the assessment was provided in 2017 by the application of one of the most popular multi-criteria TOPSIS techniques (Siksnelyte and Zavadskas, 2019).

3.2.2.1 Energy Policy Context

Ambitious targets were set by countries for their energy sectors in seeking a safe and low-carbon energy sector. The creation of a common EU electricity market is a huge job that continues after almost thirty years and it is still in the process. Electricity market development, decarbonization, improvements in infrastructure among the EU member states, and the creation of an internal EU energy market are still important questions in the EU agenda after many years. The development of the electricity sector of the EU countries is mainly based on the EU strategies (Lockwood et al., 2017). The EU legislation and various directives launched since 1990 have harmonized the electricity sector in the EU countries step by step and a common electricity policy now is being implemented for the whole region. The most important changes in the electricity sector are retail and wholesale market liberalization, production and transmission unbundling, and the establishment of an independent national regulator.

When the consequences of the energy sector to the issues of climate change was recognized, the government of the EU started to implement sustainable energy policy in the region, focusing on the implementation of targets through various strategies and directives. The EU policy for the electricity sector is supported by two pillars: Policy for market liberalization and policy for a smart and sustainable electricity sector (Pereira et al., 2018).

3.2.2.2 Indicators for Comparative Assessment of Electricity Sector Sustainability

Indicators set for the sustainability assessment of the electricity sector have been developed following methodology developed by the International Atomic Energy Agency, the International Energy Agency, the United Nations Department of Economic and Social Affairs, Eurostat and the European Environment Agency (IAEA, 2005) (EISD indicators) and based on the analysis of scientific literature (analysis of methods for sustainable energy system planning (e.g., Greening et al., 2004; Ioannou et al., 2017; Mardani et al., 2017) and impact analysis (e.g., Shortall et al., 2017), analysis of methods for renewable energy development (e.g., Cavallaro, 2009; Kurka and Blackwood, 2013; Bhowmik et al., 2017; Kumar et al., 2017), analysis of alternative methodologies for analyzing off-grid electricity supply (e.g., Bhattacharyya, 2012), and analysis of empirical studies that applied different MCDM techniques (AHP, ANP; Claudia et al., 2014; Cucchiella et al., 2017; Ligus, 2017), Fuzzy Sets (Ren and Lutzen, 2017); WSM (Klein and Whalley, 2015). The structure of EISD is presented in sub-section 3.2.1.

A set consisting of eight indicators were created to assess the sustainability of the electricity sector, taking into account the possibility of obtaining reliable and representative statistical data. Despite the fact that the classification of indicators into categories can vary, the indicators are grouped into three categories reflecting the traditional concept of sustainability and issues of the electricity sector: economic, environmental, and energy security. The created set of indicators to monitor electricity sector sustainability is presented in Table 3.16 and was applied to assess the electricity sector sustainability of EU countries.

Electricity prices (wholesale and retail) are the main economic indicators. One fundamental aspect for the electricity markets liberalization is to ensure the possibility of getting the prices as low as possible for consumers. Experience shows that to

TABLE 3.16
Set of Indicators to Monitor Electricity Sector Sustainability

Dimension	Label	Indicator	Measurement	Target value	Weight
Economic	EC1	Wholesale electricity prices	kWh/EUR, PPS, including all taxes and levies	min	1/7
	EC2	Electricity prices for households	kWh/EUR, PPS, including all taxes and levies	min	1/7
Environmental	EN1	Distribution losses	Distribution losses from final consumption, %	min	1/14
	EN2	Transformation losses	Transformation losses from gross electricity production, TOE/GWh	min	1/14
	EN3	Share of renewable energy	Share of RES in gross final electricity consumption, %	max	1/7
Energy security	ES1	Electricity interconnection	% of installed capacity	max	1/7
	ES2	Demand fulfilment with inland production	% of gross electricity production, from inland demand	max	1/7
	ES3	Import dependency from non EU countries	% of imported electricity from non EU countries from final consumption	max	1/7

Source: Created by authors.

achieve this objective is more difficult than was expected at the beginning of the liberalization process. Many countries face a variety of internal and external challenges.

The environmental dimension for assessment of the electricity sector sustainability is very important due to international commitment implementation regarding strategic environmental and climate change plans. Three environmental indicators reflecting electricity sector sustainability were chosen, which are as follows: distribution losses, transformation losses and a share of RES in gross final energy consumption. The statistics show that only about 80 percent of generated energy reaches energy consumers. A big part of energy is lost in generation, supply, and distribution activities. Therefore, distribution and transformation losses represent an important issue in measuring electricity sector sustainability from the environmental protection point of view and shows the effectiveness of activities. The share of RES shows the level of the implementation of the EU's long-term strategies, the structure of the energy mix, the climate change, and energy policy achievements.

To reflect energy security issues in evaluating the electricity sector's sustainability, the level of electricity interconnection, the fulfilment of electricity demand with inland production and percentage of imported electricity from non-EU countries were selected. The level of electricity interconnections reflects the energy security of a country. Partially, this indicator is related to the level of market competition, which allows for reaching competitive energy prices. Energy demand fulfilment with inland production is an important issue of each country's energy security. The more a country can fulfil its internal energy demand, the more energy-secure the country is. Now, EU energy policy is oriented to searching for new energy suppliers. However, the strategic long-term goal is to fulfil inland energy needs with inland production. This can be reached only by the development of new renewable energy infrastructure and by improvement in energy efficiency.

All the selected indicators for the assessment have the same weight, except distribution and transformation losses. Since they are presented separately in the statistics, their weight is divided in half. Indicators for electricity sector sustainability assessment for all EU member states are presented in Table 3.17.

TABLE 3.17
Indicators for the Electricity Sector Sustainability Assessment, 2017

Country	EC1	EC2	EN1	EN2	EN3	ES1	ES2	ES3
Austria	0.174	0.343	5.114	0.018	72.17	15.31	103.93	0.71
Belgium	0.227	0.473	4.440	0.106	17.24	18.95	99.00	0.00
Bulgaria	0.272	0.209	11.070	0.132	19.12	7.08	130.62	0.36
Croatia	0.223	0.307	10.748	0.017	46.42	51.99	65.88	22.77
Cyprus	0.240	0.273	5.022	0.125	8.9	0.00	104.73	0.00
Czech Republic	0.301	0.408	7.407	0.067	13.65	19.3	133.30	0.00
Denmark	0.226	0.277	5.201	0.000	60.36	50.57	91.51	16.85
Estonia	0.175	0.187	10.089	0.124	17.03	23.67	151.80	0.00
Finland	0.088	0.277	3.364	0.059	35.22	28.78	79.40	7.39
France	0.160	0.275	8.703	0.134	19.91	9.44	114.56	1.45
Germany	0.265	0.446	5.086	0.073	34.41	8.95	117.08	0.00
Greece	0.243	0.252	1.981	0.066	24.47	10.6	97.44	8.98
Hungary	0.225	0.218	8.718	0.083	7.49	58.25	75.47	12.28
Ireland	0.196	0.379	8.379	0.065	30.09	7.41	106.41	0.00
Italy	0.273	0.315	6.184	0.052	34.1	8.18	92.29	7.15
Latvia	0.328	0.276	7.315	0.020	54.36	23.67	108.22	14.72
Lithuania	0.224	0.184	8.341	0.030	18.25	23.67	35.27	40.49
Luxembourg	0.118	0.224	2.438	0.012	8.05	109.22	34.13	0.00
Malta	0.251	0.446	5.718	0.095	6.58	24.24	66.93	0.00
The Netherlands	0.178	0.325	4.858	0.048	13.8	18.11	100.25	4.47
Poland	0.331	0.333	6.827	0.003	13.09	4.05	108.98	0.61
Portugal	0.309	0.487	10.702	0.061	54.17	8.73	112.64	0.00
Romania	0.228	0.261	14.314	0.074	41.63	6.92	114.82	2.54
Slovakia	0.345	0.346	4.675	0.027	21.34	43.29	98.09	0.03
Slovenia	0.198	0.312	6.557	0.058	32.43	83.56	112.47	0.00
Spain	0.344	0.633	11.193	0.092	36.34	5.79	101.97	0.00
Sweden	0.132	0.291	6.496	0.065	65.89	25.61	118.22	6.97
United Kingdom	0.167	0.253	8.491	0.077	28.11	5.00	101.23	0.00

Source: Created by authors based on data from European Commission (2022b) and Eurostat (2022).

3.2.2.3 MCDM Tool

The application of various multi-criteria analysis techniques to solve energy sustainability issues started to be a very popular tool due to plenty of different approaches, the scientific justification, and accuracy of calculations. According to the literature analysis, the most applied MCDM techniques to solve energy sustainability issues were AHP, TOPSIS, PROMETHEE, ELECTRE methods (Wang et al., 2011; Mardani et al., 2017; Siksnelyte et al., 2018). To assess electricity sector sustainability of the EU member states, the multi-criteria evaluation TOPSIS technique was applied for calculations and ranking.

The TOPSIS technique was introduced in 1980 by Hwang and Yoon (Hwang and Yoon, 1981). These scientists developed a methodology to rank alternatives following the concept that the best alternative is the closest distance from the best solution and the largest distance from the worst solution. Each criterion of the alternative has a tendency for monotonically increasing or decreasing utility. Therefore, the ideal and opposite to ideal solutions can be easily determined.

The TOPSIS technique has been applied in much research on energy sustainability measurement and energy planning (e.g., Afsordegan et al., 2016; Ozcan et al., 2017; Vavrek and Chovancova, 2019; Solangi et al., 2019). The method can be characterized as having low computational burden, and the theoretical concept of the ideal solutions is followed. The approach used is reasonable for comparison in the international level, where countries can be compared to the ideal solutions acting as the data-driven benchmarks. The vector normalization is also advantageous, because it considers all the values observed when normalizing a certain criterion. The ranking of the EU member states under this method is calculated in accordance with the following seven key steps:

Step 1. Decision matrix creation with m alternatives and n criteria:

$$D = \left[x_{ij} \right] = \begin{matrix} a_1 \\ a_2 \\ \cdots \\ a_m \end{matrix} \begin{bmatrix} X_1 & X_2 & \cdots & X_n \\ x_{11} & x_{12} & \cdots & x_{1n} \\ x_{21} & x_{22} & \cdots & x_{2n} \\ \cdots & \cdots & \cdots & \cdots \\ x_{m1} & x_{m2} & \cdots & x_{mn} \end{bmatrix} \qquad (3.12)$$

Step 2. The normalized matrix is obtained by equation 3.13:

$$\overline{x}_{ij} = \frac{x_{ij}}{\sqrt{\sum\limits_{i=1}^{m} x_{ij}^2}} ; i = \overline{1, m}, . j = \overline{1, n}, \qquad (3.13)$$

Step 3. The weighted normalized matrix calculation:

$$V = \begin{bmatrix} w_1 r_{11} & w_2 r_{12} & \cdots & w_n r_{1n} \\ w_1 r_{21} & w_2 r_{22} & \cdots & w_n r_{2n} \\ \cdots & \cdots & \cdots & \cdots \\ w_1 r_{m1} & w_2 r_{m2} & \cdots & w_n r_{mn} \end{bmatrix}; \sum_{j=1}^{n} w_j = 1 \qquad (3.14)$$

Step 4. The determination of positive A^+ and negative A^- ideal solutions:

$$A+ = \{maxjvij \mid i \in I), (minjvij \mid i \in I'), j = \overline{1,n},\} = \{v^+1, v^+2 \ldots v^+n\}; \qquad (3.15)$$

$$A- = \{minjvij \mid i \in I), (maxjvij \mid i \in I'), j = \overline{1,n},\} = \{v^-1, v^-2 \ldots v^-n\} \qquad (3.16)$$

Step 5. Measurement of the relative distance of each positive $A+$ and negative $A-$ ideal solution:

$$S_i^+ = \sqrt{\sum_{j=1}^{n} \left(v_{ij} - v_j^+\right)^2}, j = \overline{1,m}, \qquad (3.17)$$

$$S_i^- = \sqrt{\sum_{j=1}^{n} \left(v_{ij} - v_j^-\right)^2}, i = \overline{1,m}, \qquad (3.18)$$

Step 6. Determination of the relative closeness of each alternative to the ideal solution:

$$C_i = \frac{S_i^+}{S_i^+ + S_i^-}; \qquad (3.19)$$

Step 7. Identification of the best alternative, that is, the identification of the alternative that is closest to 1. The more Ci value is close to 1, the more the alternative is close to the ideal solution A+, that is, the best alternative under assessment will be the one with the highest Ci value. Based on the values of Ci, a series of priorities of alternatives is created.

3.2.2.4 Multi-Criteria Assessment Results

The results of the electricity sector sustainability assessment among EU member states by the application of the TOPSIS technique are presented in Table 3.18.

According to the results, the most sustainable electricity sector is in Slovenia (0.7515), and Luxembourg (0.7091) took second place in the ranking. Although, the prices of electricity for households were not very low in Slovenia, the country can

TABLE 3.18
The Results of the Assessment of EU Countries' Electricity
Sector Sustainability, 2017

Alternatives	S_i^+	S_i^-	C_i	Rank
Slovenia	0.0432	0.1307	0.7515	1
Luxembourg	0.0587	0.1431	0.7091	2
Austria	0.0747	0.1231	0.6224	3
Slovakia	0.0729	0.1165	0.6149	4
Sweden	0.0692	0.1100	0.6139	5
Estonia	0.0829	0.1181	0.5874	6
Denmark	0.0674	0.0949	0.5845	7
Finland	0.0740	0.1029	0.5817	8
Portugal	0.0879	0.1144	0.5654	9
United Kingdom	0.0899	0.1141	0.5594	10
Germany	0.0884	0.1121	0.5590	11
Ireland	0.0888	0.1121	0.5581	12
Czech Republic	0.0889	0.1116	0.5567	13
Romania	0.0878	0.1093	0.5546	14
Belgium	0.0888	0.1101	0.5534	15
Hungary	0.0776	0.0940	0.5478	16
France	0.0915	0.1099	0.5457	17
Malta	0.0916	0.1092	0.5439	18
The Netherlands	0.0866	0.1031	0.5434	19
Bulgaria	0.0950	0.1127	0.5426	20
Poland	0.0980	0.1109	0.5309	21
Spain	0.0991	0.1088	0.5235	22
Cyprus	0.1018	0.1117	0.5232	23
Italy	0.0898	0.0962	0.5170	24
Latvia	0.0831	0.0888	0.5167	25
Greece	0.0907	0.0939	0.5086	26
Croatia	0.0826	0.0777	0.4845	27
Lithuania	0.1349	0.0485	0.2644	28

Source: Created by authors.

be characterized by a high-level of electricity network interconnection, the fulfilment of electricity demand with inland energy generated, and independency of electricity imports produced outside the EU. Luxembourg is notable for low electricity prices and high efficiency in electricity distribution and transformation, and for a high level of electricity networks interconnection. The country did not import electricity from non-EU countries. Austria took third place in the final ranking (0.6224), Slovakia was the fourth (0.6149), Sweden took fifth (0.6139).

Lithuania (0.2644) was last in the ranking in 2017. Also, it should be noted, the score of Lithuania in the assessment according to the criteria selected was very far from Croatia (0.4845), which took the penultimate place. The main cause for such results is a high dependency on imported electricity from non-EU countries, where 40.5 percent of final electricity consumed came into Lithuania from Russia (14.8%) and Belarus (25.7%), 22.8 percent of final electricity consumed in Croatia came from Bosnia and Herzegovina (17.7%) and Serbia (5.1%). The results of the assessment of all other EU member states from 6th to the 26th place did not differ very significantly and get in the range (0.5874 – 0.5086).

Sustainability issues in the electricity sector are very important for each country's energy system. This is one of the main sectors, which can help to achieve climate change and energy policy goals. In the EU, the biggest portion of energy is consumed by meeting heating and cooling needs. However, the biggest share of energy goes for heating. Therefore, the decisions regarding heating and cooling decarbonization can have an essential impact on further electricity sector development. For example, the electrification of heating and cooling systems can double the electricity demand (Connolly, 2017). There are also other factors due to which the electricity demand can increase significantly. For example, the switch to electric vehicles will boost electricity demand (Strielkowski et al., 2019), and so forth. Definitely, the development of new renewable electricity infrastructure will have a major impact on the overall energy sector sustainability in future years. The share of RES in the electricity mix is still too low in many EU countries. Despite the fact, that all the EU countries, except France, implemented the strategy Europe 2020 national target for RES in 2020, and for some countries the national targets set were very low. Also, it can be noticed that there is still a lack of initiatives to encourage energy consumers to switch to sustainable technologies. In some countries, there is also insufficient information about support opportunities, possible benefits, advantages of technologies, and so forth. For example, in Greece the results of the study by Manolopoulosa et al. (2016) showed that bureaucracy (long administrative procedures, disagreement regarding responsibilities among national institutions), government elections (slow down in the processes before and after the change of government, possible changes in procedures), low public acceptance (slow down by delays the process of new project implementation) are the main barriers determining the slow development of new RES electricity infrastructure. It can be assumed that in other countries having high level of bureaucracy similar problems are faced regarding the development of the RES electricity sector. The countries should find the most effective measurement for each group of energy consumers to encourage the development of RES. The experience regarding subsidies for RES electricity is widely criticized in scientific literature and many empirical studies for high costs and a comparatively low level of benefits (Chaton and Guilliminet, 2013; Flues et al., 2014; Del Río et al., 2017; Lehmann et al., 2019).

The EU countries' electricity sector sustainability assessment showed the most sustainable electricity sectors in 2017 were in Slovenia and Luxembourg. Lithuania and Croatia took the two last positions in the ranking mainly due to their dependence on imported electricity from non-EU countries. Although, the EU climate change and energy policy targets set by the Europe 2020 strategy have been achieved, much more effort toward energy sustainability should be expended. For some countries, the targets set were quite low; also the analysis of various scientific and empirical studies showed that a lot of problems regarding bureaucracy, public acceptance, technologies promotion, support mechanisms, and so forth. were not solved. Also, a very important issue for the whole region is dependency on imported energy, especially from Russia. Now, one of the most crucial issues for the region is to find alternative suppliers as soon as possible and to increase RES in final energy consumption, as well energy efficiency. The proposed framework for the evaluation covers three dimensions critical for the energy sector – economical, environmental, and energy security. This framework can be applied to monitor electricity sector sustainability in the future, following the progress and efficiency of the various policy measures applied.

3.2.3 FRAMEWORK TO MONITOR HEATING SECTOR SUSTAINABILITY

This subsection presents a framework heating sector sustainability assessment. The developed framework was applied for the comparative assessment of North European countries in 2018 (Siksnelyte-Butkiene et al., 2021). One of the most popular multi-criteria technique TOPSIS was used for calculations and ranking. Also, the sensitivity analysis was performed by creating four different weighting schemes.

3.2.3.1 Energy Policy Context and Overview of EU Heating Sector

Despite the fact that in the EU half of the energy consumed is for heating and cooling; the sector can not be described by a significant share of renewable energy, high absorption of innovative technologies, or a high level of energy efficiency. There are a lot of issues requiring improvements in energy generation, supply, and consumption. For example, fossil fuels account for about 75 percent of the heating and cooling energy mix. Especially, a significant amount of energy is generated from gas, and this causes a lot of energy security problems in the region.

The main instrument to develop a more sustainable EU heating sector is constituted by various strategies and directives, which guide EU countries and give policy directions and goals they are committed to implement. In 2016, the Heating and Cooling Strategy (European Commission, 2016c) was adopted with the aim of strengthening energy security in the region and achieving the 2030 and 2050 climate change and energy targets. The strategy focuses on energy efficiency issues through building renovation, absorption of efficient technologies, and increasing the district heating share. Although this strategy is the first directly targeted initiative to take regional action in the sector, the previous effort to achieve climate change and energy policy objectives can also be referred to as having an impact on the decarbonization and improvements in efficiency of the sector.

The objectives regarding climate change and energy policy are defined in different strategies. One of the most important was the ten years' strategy Europe 2020, which aimed to reduce GHG emissions by 20 percent compared to the emissions level in 1990, 20 percent of energy in final energy consumption should be generated from RES, and increase 20 percent in energy efficiency. Also, several other strategies were adopted to encourage the implementation of these targets. For instance, the Energy Roadmap (European Commission, 2011) presented the targets of the EU energy sector decarbonization and singled out the key steps to achieve them. The

Energy Efficiency Directive (2012/27/EU) (European Parliament and Council of the European Union, 2012) introduced the measures for the energy efficiency target achievement by 2020. The strategy Energy 2030 (European Commission, 2014b) upgraded the climate change and energy targets and is a significant step toward achieving the 2050 energy and climate change targets. According to the strategy, GHG emissions should be reduced by 40 percent compared to the level in 1990, the portion of RES in gross final energy consumption should reach 27 percent and increase energy efficiency by 25 percent until 2030. Also the strategy stimulates diversification of energy sources, market competitiveness and the development of innovative and efficient energy technologies. An other important document is the Energy Union Package (European Commission, 2015), which seeks to increase energy efficiency in buildings and encourages the investments in innovative low-carbon energy technologies. The attention to the buildings efficiency problems, energy generation technologies used, and the portion of RES in the final energy balance was also drawn in the Clean Energy for all Europe package (European Commission, 2016a). The Energy Efficiency Directive (2018/2002) (European Parliament and Council of the European Union, 2018a) obligated to implement further initiatives to increase energy efficiency in heating and cooling sectors. The renewed RES Directive (2018/2001) makes it obligatory to determine national targets and create national action plans for the development of RES. According to the directive, each EU country should increase the share of RES in heating and cooling for about 1.3 percent annually by 2030 (European Parliament and Council of the European Union, 2018b).

Costs of heating vary across the EU countries and depend on many aspects, such as: market regulation, legal basis, energy mix, social policy measures, and so forth. Figure 3.1. provides composition of energy sources for heating in the EU in the years 1990–2020.

The statistical data of three decades show the progress in reducing consumption of fossil fuels for heating generation, but the progress is still not big enough to fight

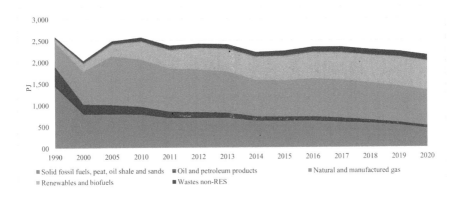

FIGURE 3.1   The energy mix in gross heat generation in the EU, PJ/years.

Source: Based on data from Eurostat (2022).

against climate change and other energy problems. Renewables and biofuels accounts 32 percent in gross heat generation mix, wastes non-RES only 6 percent in 2020. Accordingly, 62 percent of heat is generated from fossil fuels. It should be noted that the structure of energy sources was changing during the last three decades: The use of coal was significantly reduced, but the heat generated from natural gas doubled. Countries with a high share of RES are more energy efficient. For example, energy efficiency of heating supply in Finland and Sweden is more than 90 percent. These countries are notable for a high share of heat supplied via district heating, electricity, or renewables. The EU heating sector relies on gas, and this is not only an environmental problem, but also an energy security issue, as mentioned before. The development of new renewable infrastructure and energy savings are very important components, ensuring the implementation of climate change and energy goals. And the heating sector has a huge potential for improvement. These improvements can help to achieve energy independence and to ensure security of energy supply. The development of innovative and efficient technologies can help to fight against climate change and energy poverty issues.

The different climatic conditions of the EU countries accordingly determine the share of expenditures for heating and cooling. Based on that, the targets for climate change and energy are focused on the sector that accounts for the largest part in energy consumption structure. Figure 3.2 presents the portion of RES in heating and in heating and cooling in the EU member states.

Several countries can be singled out as having significant achievements in terms of renewable energy usage in the heating and cooling sector. For example, in Sweden, RES in heating account for 76 percent and RES in cooling and heating account for 65 percent in 2018. More than half of energy for heating is generated from biofuels in Estonia, Denmark, Lithuania, and Luxembourg. However, despite these exemptions,

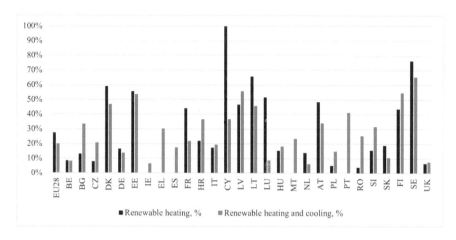

FIGURE 3.2   Share of RES in heating and in heating and cooling in 2018, %.

Note: Abbreviations of the EU countries are provided in Appendix 3.1.

Source: Based on data from Eurostat (2022).

the fuel mix of the remaining countries mainly consists of fossil fuels. Another relevant indicator, which reflects countries' achievements in shaping the sustainable energy sector, is reduction of GHG emissions. Figure 3.3. and Figure 3.4 present the main indicators for countries' comparison regarding this aspect:

The statistical data proves that the countries with the highest portions of solid fossil fuels in their energy mix have the highest level of GHG emissions, both in terms of GDP and population. As an example, Estonia stands out for an exceptionally large amount of GHG emissions, and although energy generated from solid fossil fuels accounts for only 12.56 percent of heating energy generated, the main fuel in electricity generation mix is solid fossil fuels, accounting for more than 75 percent.

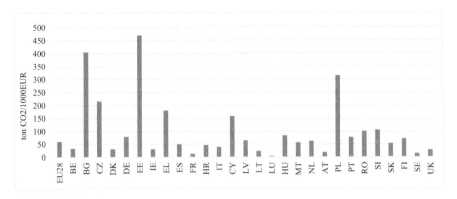

FIGURE 3.3   GHG emissions per GDP in the EU member states in 2018, ton CO_2/1000EUR.

Note: Abbreviations of the EU countries are provided in Appendix 3.1.

Source: Based on data from Eurostat (2022).

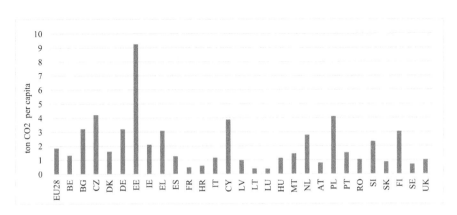

FIGURE 3.4   GHG emissions per capita in the EU member states in 2018, ton CO_2.

Note: Abbreviations of the EU countries are provided in Appendix 3.1.

Source: Based on data from Eurostat (2022).

Another example is Poland, where coal is the main fuel for heating and electricity generation, accounting for about 80 percent in final energy mix. The amount of GHG emissions reflect not only the issues of climate change, but also people's health, well-being, and living conditions. Therefore, decarbonization of the energy sector and also the decarbonization of whole economy is crucial issue nowadays. From this also rises the necessity to measure sustainability of the energy sector and the efficiency of applied development measures to implement the objectives set.

3.2.3.2 Indicators for Comparative Assessment of Heating Sector Sustainability

The scientific literature is rich with numerous evaluation systems developed and various calculation methodologies proposed (Stanujkic et al, 2019; Liu et al., 2019; Tabatabaei et al., 2019) to assess different issues of sustainability in the heating sector. Commonly, different heating technologies are evaluated for district heating (Mazhar et al., 2018; Pinto et al., 2019; Kirppu et al., 2018) or for residential buildings (Yang et al., 2018; Zhang et al., 2019a; L. Saleem and Ulfat, 2019; Seddiki and Bennadji, 2019), or projects (Herrera et al., 2020). However, there are only a few studies evaluating the sustainability of the heating sector as a whole. Table 3.19 provides the summary of studies evaluating the sustainability of the heating sector and indicators used for the measurement:

TABLE 3.19
Summary of Indicators Used to Assess Heating Sector Sustainability

Reference	Method applied	Dimension	Indicators
Liposcak et al. (2006)	MCDM technique – the Analysis and Synthesis Parameters under Information Deficiency (ASPID)	Economic	Investments; fuel cost
		Environmental	CO_2, SO_2 and particulate matter (PM) emissions
		Social	Health; public acceptance
Kuznecova et al. (2017)	-	Economic	GDP; % of low-income households; % of costs from income; energy costs for one inhabitant
		Environmental	% of RES; % of fossil fuels; CO_2 and PM emissions; heating consumption in household
		Social	Quantity of rooms in a dwelling; quantity of rooms per one household member; dwelling size; expenditure problems; environmental problems

TABLE 3.19 (Continued)
Summary of Indicators Used to Assess Heating Sector Sustainability

Reference	Method applied	Dimension	Indicators
Kveselis et al. (2017)	Life Cycle Assessment (LCA)	Environmental	CO_2 emissions; renewable and recycled energy; primary energy
Alola et al. (2019)	Autoregressive Distributed Lag	Economic	GDP; population
		Environmental	Consumption of energy generated from fossil fuels; ecological footprint
Jovanovic et al. (2019)	MCDM technique – ASPID	Economic	Energy consumption; energy generation (by GDP); total costs for heating; energy costs for one inhabitant; potential to reduce consumption of gas; potential to increase share of RES, potential to reduce gas cost
		Environmental	CO, SiO, NO and NMVOC emissions
		Social	Security of energy supply; energy consumption; % of RES; % of income for heating needs
Hehenberger-Risse et al. (2019)	Impact assessment	Economic	Regional added value; heating price
		Environmental	Renewable energy; non-renewable energy; overall energy efficiency; environmental impact; $CO2, SO2$ emissions; wastewater; area;
Hobley (2019)	Scenario analysis	Environmental	CO_2 emissions
Rutz et al. (2019)	MCDM approach, scenario analysis	Economic	Investments; fuel cost
		Environmental	CO_2, SO_2, NO_x and PM emissions
		Social	Employment rate increase; income generation in local-level; regional development
Chen et al. (2020)	MCDM approach, scenario analysis	Economic	Investment cost; cost saving ratio
		Environmental	CO_2, SO_2, NO_x and PM emissions
		Social	Employment opportunities
		Energy	Heating output; energy saving ratio; electricity consumption from grid; consumption of solar energy

Source: Created by authors.

There are studies that have attempted to assess the sustainability of the heating sector based on the traditional concept of sustainable development. For example, Liposcak et al. (2006) used three economic, three environmental, and two social indicators for the assessment of heating sector sustainability in Croatia. The authors applied the ASPID method for the calculations and presented three scenarios for heating sector development in the future. Different Serbia district heating scenarios were analyzed by Jovanovic et al. (2019) by application of the traditional approach to sustainability. The authors created a framework from six economic, four environmental, and four social indicators, representing the aspects and consequences of sustainable heat production and consumption. The multi-criteria ASPID technique was applied for calculations. The analysis of district heating systems and measurements of sustainability toward transition to renewable energy based heating sectors in five Southeast European countries was performed by Rutz et al. (2019). The ten indicators framework, reflecting the traditional approach of sustainability was used (three economic, four environmental, and three social indicators). For the assessment of sustainability of district heating systems, Chen et al. (2020) singled out indicators, which reflect consumption of energy, into a separate category (Energy indicators). However, these indicators are very similar to those measured by Rutz et al. (2019) and can be assigned to the traditional approach. Kuznecova et al. (2017) presented an index for the sustainability assessment of heat energy generation in the household sector. The index reflects traditional concepts of sustainability and consist of 4 economic, 22 environmental and 7 social criteria.

Other studies focus on environmental aspects in the heating sector. For example, Kveselis et al. (2017) analyzed the green energy labelling scheme for district heating and cooling systems in Lithuania. The authors performed LCA and measured sustainability only on three environmental criteria (CO_2 emissions, primary energy, renewable and recycled energy). Hobley (2019) performed scenario analysis for the decarbonization of the heating sector in the UK with the aim to ensure sustainability objectives, security of energy supply, affordable price, and adaptation of new technologies. The authors expressed heating sector sustainability only through the measurement of CO_2 emissions. Other studies sought to combine environmental and economic aspects. For example, Alola et al. (2019) studied the impact of macroeconomic variables to the environmental sustainability in the heating and cooling sector of the United States. The study takes into account six indicators, which can be assigned to economic and environmental dimensions. Hehenberger-Risse et al. (2019) presented a framework for the measurement of the environmental impact of heating supply systems. The framework consists of twelve indicators, which reflect environmental and economical aspects.

The essential social aspects of the heating sector are analyzed in energy poverty studies. Numerous different indicators are used, and various indexes are created to measure energy poverty (Siksnelyte-Butkiene et al., 2021a). The extent of energy poverty depends on many factors, which include not only external conditions, based on geographical specifics or national energy system characteristics, but also on individual characteristics of households, such as disposable income and habits (Pino-Mejias et al., 2018; Casquero-Modrego and Goni-Modrego, 2019; Okushima, 2019;

Recalde et al., 2019), affordability of innovative technologies (Ayodele et al., 2015), efficiency of dwellings (Gouveia et al., 2019; Castano-Rosa et al., 2020), and so forth. In general, indicators on demographic characteristics, income, habits of inhabitants, and the level of energy efficiency reflects the social aspects of energy poverty. For example, in the EU, several self-reported EU surveys on income and living condition (EU-SILC) indicators are widely used in energy poverty studies to reflect access to energy services, the energy efficiency level of a dwelling, and the level of thermal comfort. Therefore, these indicators are suitable to measure the social aspects of the heating sector.

The created framework assessing heating sector sustainability issues are applied for the evaluation of seven Northern European countries: Denmark, Finland, Sweden, the Baltic countries (Estonia, Latvia, Lithuania) and the UK. These countries have been selected taking into account the geographical conditions, which are very important factors when considering heating issues (Bellos and Tzivanidis, 2017; de Rubeis et al., 2020). Due to the location and climatic conditions of the selected countries, the geographical aspect plays an important role, and the heating sector is one of the most important. Also, for heating in these countries, gas is a very significant energy source together with renewables. A set of nine indicators reflecting traditional concepts of sustainability has been developed. The TOPSIS technique has been selected for evaluation and ranking of the countries. Also, a sensitivity analysis was performed by creating four scenarios.

The measurement of heating sector sustainability is important, not only in scientific terms, but also at the practical level. It allows for measuring the efficiency of the measures implemented, justifies political decisions, and allows for designing guidelines for improvement. The set of indicators for the heating sector sustainability assessment is presented in Table 3.20.

Energy cost, GHG emissions (Budzianowski, 2012; Bilgili, 2012; Goodbody et al., 2013), RES fraction (Kveselis et al., 2017), energy efficiency (Harmsen et al., 2011; Sperling and Moller, 2012), and living conditions of energy consumers (Yılmaz and Selim, 2013) are the fundamental components of sustainable heating and also of the whole energy sector functioning.

In the selected countries under assessment, the dominated energy source for heating generation is gas in both households and public sectors. Therefore, the gas price for households' and non-households' consumers (EC1 and EC2) reflect the heating costs.

The application of financial measures in order to modernize the sector (EC3) shows whether solutions are being sought at the national level. New technologies require high investment costs and public acceptance. Therefore, the application of financial mechanisms is an important aspect to encourage private investments in new technologies (Zhang et al., 2019b; Rodriguez-Alvarez et al., 2019; Kyprianou et al., 2019) and to solve issues of energy efficiency (Masini and Menichetti, 2013; Lukanov and Krieger, 2019). The existence of financial measures covers both initiatives to modernize household technologies and infrastructure projects, as well as encourages energy efficiency savings in the heating sector.

The percentage of people who live in buildings with leakage (SO1) represents the social dimension and also indicates the efficiency of buildings. Arrears on utility bills

TABLE 3.20
Set of Indicators for the Heating Sector Sustainability Assessment

Dimension	Label	Indicator	Measurement	Target	Data source
Economic	EC1.1	Gas price for	kWh/EUR, PPS,	min	Eurostat
	EC1.2	households*	including all	min	
	EC1.3		taxes and levies	min	
	EC2.1	Gas price for	kWh/EUR, PPS,	min	Eurostat
	EC2.2	non-household	including all	min	
	EC2.3	consumers**	taxes and levies	min	
	EC3	The existence of financial measures for the heating sector modernization	Binary scale	max	National Energy and Climate Plans
Environmental	EN1	RES and biofuels for heat generation	% of generated heat	max	Eurostat
	EN2	Waste non-RES for heat generation	% of generated heat	max	Eurostat
	EN3	GHG emissions (electricity and heat)	ton/per capita	min	Eurostat
Social	SO1	People living in a houses with leakages	% of total population	min	Eurostat
	SO2	People having arrears on utility bills	% of total population	min	Eurostat
	SO3	People unable to heat home adequately	% of total population	min	Eurostat

Source: Created by authors.

* EC1.1 (Consumption < 20 GJ), EC1.2 (Consumption < 200 GJ), EC1.3 (Consumption > 200 GJ).

** EC2.1 (Consumption < 1000 GJ), EC2.2 (Consumption < 10000 GJ), EC2.3 (Consumption <100000 GJ).

(SO2) indicates the efficiency of national social and energy policy implemented. The percentage of people who are unable heat homes adequately warm (SO3) reveals energy poverty among people and the necessity of political intervention to solve the problem. These three self reporting indicators are part of EU-SILC (European Commission, 2014c).

Renewable energy development is a crucial aspect that can help shift the energy system toward sustainability (Dombi et al., 2014). Heat generation from RES and biofuels (EN1) is a priority indicator, both in the national and whole EU levels. It was known that the specifics of the heating sector allow for reaching the highest efficiency of RES implementation (IEA, 2012). The development of renewable infra-structure in heating is related to many positive aspects such as: development of new effective energy technologies; increases in energy efficiency (Kveselis et al., 2017);

TABLE 3.21
Supporting Indicators

Country	Gross heat generated, PJ	Heat generation from waste non-RES, PJ	Heat generation from RES and biofuels, PJ	GHG emissions (electricity and heat), mio ton CO_2	Population, thousands
Denmark	130.8	12.7	77.3	9.1	5781
Estonia	24.30	0.9	13.5	12.2	1319
Finland	170.8	5.6	74.7	16.4	5513
Latvia	29.7	0.0	13.9	1.8	1934
Lithuania	34.6	0.7	22.8	1.0	2809
Sweden	185.8	21.8	141.9	6.7	10120
UK	66.3	0.4	4.4	66.3	66274

Source: Created by authors.

improvement in environmental, and people's health issues (Yadoo and Cruickshank, 2012); strengthening energy security; and fighting against energy poverty (Sovacool and Mukherjee, 2011; Sovacool, 2011; Okushima, 2019). However, despite the advantages of RES usage in heating, the sector relies on fossil fuels, especially on coal and natural gas in many countries. Heat generated from waste non-RES (EN2) shows other significant problems, where the heating sector can contribute (Maghmoumi et al., 2020). Waste incineration is a big challenge facing the heating sector (Nami et al., 2019). The level of GHG emissions (EN3) reflects the quality of the environment. Also the indicator is strongly related to human health and well-being issues (Ekholm et al., 2014; Dziugaite-Tumeniene et al., 2017).

The set of indicators was created with the aim not only to select representative indicators reflecting the goal of the evaluation, but also to ensure the repeatability of the study. Therefore, all indicators are comparable, publicly available, and can be easily found. However, the environmental dimension indicators were calculated using supporting indicators (Table 3.21) used to calculate a share of RES and biofuels (EN1), share waste non-RES (EN2) and GHG emissions in electricity and heating (EN3).

3.2.3.3 MCDM Implementation

As mentioned before, MCDM methods offer the possibility to evaluate many conflicting aspects (Bhardwaj et al., 2019) of energy sector sustainability and to determine which alternative is the best according to the criteria selected (Si et al., 2016). In order to measure heating sector sustainability and to rank the selected countries according to their achievements, the multi-criteria evaluation TOPSIS technique (Hwang and Yoon, 1981) was used for calculations. The main steps of the TOPSIS technique is presented in sub-section 3.2.2.

For the sensitivity analysis four scenarios were created that describe the unique importance of sustainability issues (Table 3.22). The scenarios were created for the analysis of changes in the ranking due to the uniqueness of each sustainability

TABLE 3.22
Weighting Schemes for Sensitivity Analysis

Scenario	Economic dimension	Social dimension	Environmental dimension
Scenario 1 (S1)	0.(3)	0.(3)	0.(3)
Scenario 2 (S2)	0.5	0.25	0.25
Scenario 3 (S3)	0.25	0.5	0.25
Scenario 4 (S4)	0.25	0.25	0.5

Source: Created by authors.

dimension. Each dimension under assessment was attributed with intuitively chosen weights verified by experts in the field. The weights applied are reasonable in that no dimension is ignored, as the minimum limit is 0.25, whereas excessive importance is also prevented by using the maximum limit of 0.5.

All three dimensions are considered as equal in the S1, and therefore equal weights in the assessment are given for each dimension. In S2, the most significant is the economic dimension; in S3, social, and in S4, environmental.

The weights among indicators in economic and social dimensions are equal; in the environmental dimension half of the total weight is dedicated to RES while the other two indicators are divided equally. Since the gas price (EC1 and EC2) varies depending on the consumption level, prices for different consumption levels are included, and the weights of these indicators are distributed to sub-indicators accordingly.

3.2.3.4   Multi-Criteria Assessment Results

As an example, the assessment was performed for the analysis of countries in 2018. The data is provided in Table 3.23. According to the data, the best performing countries by analyzing only environmental indicators were Sweden and Denmark. In terms of indicators selected to reflect social dimension, the best results were reached in Finland and Sweden. The most preferable values of economic indicators were in the UK and Estonia.

The assessment of heating sector sustainability was calculated, and countries under assessment were ranked by application of the TOPSIS technique. The sensitivity analysis for four scenarios was performed and a summary of the results is presented in Table 3.24. The sensitivity analysis showed that, despite the different weighting schemes, the best performing countries in terms of sustainable heating sector development are Sweden, Denmark, and Finland. These countries did not change their positions and took the first three places in the ranking. Only the changes in weighting of environmental aspects showed variations in the ranking of other countries. However, the three best performing countries remain in the same positions in all weighting scenarios.

Sweden took first place in the ranking, Denmark the second, and third was Finland. These three best performing countries did not change their positions in the ranking despite the different weighting schemes. Therefore, it can be assumed that results are reliable.

TABLE 3.23
Data for the Assessment, 2018

Country	Economic dimension							Environmental dimension			Social dimension		
	EC1.1	EC1.2	EC1.3	EC2.1	EC2.2	EC2.3	EC3	EN1	EN2	EN3	SO1	SO2	SO3
Denmark	0.081	0.069	0.066	0.066	0.064	0.056	1	59.10	9.71	1.58	16.4	5.1	3
Estonia	0.064	0.055	0.052	0.053	0.052	0.051	1	55.56	3.70	9.23	13.6	6.5	2.3
Finland*	N/A	N/A	N/A	0.066	0.061	0.059	1	43.74	3.28	2.98	4.6	7.7	1.7
Latvia	0.100	0.064	0.064	0.069	0.062	0.056	1	46.80	0.00	0.95	23.5	11.6	7.5
Lithuania	0.103	0.063	0.050	0.082	0.078	0.072	1	65.90	2.02	0.36	14.8	9.2	27.9
Sweden	0.160	0.101	0.101	0.089	0.079	0.069	1	76.37	11.73	0.66	7.8	2.2	2.3
UK	0.071	0.047	0.041	0.052	0.034	0.030	1	6.64	0.60	1.00	17.6	5.4	5.4

Source: Created by authors.

* Households do not consume natural gas for heating purposes in Finland, therefore only the gas price for non-household consumers was accounted and the weight for Finland of EC2 indicator was doubled.

TABLE 3.24
The Results of the Assessment, 2018

Countries	S1		S2		S3		S4	
	C_i	Rank	C_i	Rank	C_i	Rank	C_i	Rank
Sweden	0.855	1	0.755	1	0.890	1	0.906	1
Denmark	0.758	2	0.744	2	0.743	2	0.775	2
Finland	0.657	3	0.656	3	0.728	3	0.612	3
Estonia	0.560	4	0.570	4	0.657	4	0.501	6
UK	0.523	5	0.541	5	0.604	5	0.476	7
Latvia	0.520	6	0.521	6	0.504	6	0.529	5
Lithuania	0.465	7	0.466	7	0.341	7	0.535	4

Source: Created by authors.

Mainly, the leadership in the development sustainable heating sector have shown significant achievement during the last decades in development of renewable energy infrastructure. For example, in the leading country, Sweden, more than 75 percent of heat demand is satisfied by renewables (including biomass), of which 8 percent is ambient heat. Sweden is distinguished from all EU member states by the largest usage of ambient heat, generating about 90 percent of all EU ambient heat. Also, in Sweden the significant part of energy for heating is generated from waste non-RES, accounting for 17 percent of total quantity of incinerated non-RES wastes for heating in the whole EU. Sweden has applied various financial mechanisms to increase energy efficiency in buildings (renovation), to increase development of photovoltaic cells and solar heat, and to support technologies for self-generated electricity storage (Regeringen, 2020). The country is also distinguished with very positive social indicators. In 2015, only 2.2 percent of the people in Sweden faced economic problems in paying their utility bills; only 2.3 percent did not heat their home enough, and 7.8 percent of the population lived in dwellings with leakages.

Denmark, which took second place in the ranking, also meets almost 60 percent of its heating needs by RES. The significant portion of the biomass for heat generation occurs in large scale combined heat and power generation at centralized plants converted from coal based production. It was mainly empowered by the incentives for biomass conversions enacted with the Energy Agreement from 2012. In homes that do not have access to district heating or natural gas, bioenergy often was the most convenient and cheapest way to heat. Natural gas is the second most popular way to heat their homes, accounting for about 15 percent of the market in 2018. In the final heating mix natural gas accounted for 17 percent, hard coal, 12 percent, electricity, 4.5 percent. Denmark seeks to create a net zero emissions energy system by 2050. Also, it should be highlighted that 10 percent of heating was generated using waste non-RES incineration, and it accounted for 10 percent of total incinerated non-RES wastes for heating in the EU. In terms of social indicators, the indicator reflecting energy efficiency issues (SO1) was quite low. Despite that, two other social indicators reflect adequate heating (SO3), and the people's ability to pay energy bills (SO2) was very good. As mentioned, the development of district heating systems in the country allows for more quickly reaching sustainability in the sector. The district heating meets about two thirds of heating needs in Denmark. An analysis of the energy mix data showed a clear trend toward reducing fossil fuels usage and increasing the usage of biofuels. The country seeks that 90 percent or more of energy in district heating would be based on non-fossil fuels by 2030. It is expected that by 2030, almost 80 percent of energy used for district heating will be from RES (especially from biomass and heat pumps). It should be noted, that the reason RES would not be more than 80 percent is because of the high share of non-RES waste incineration (Danish Ministry of Climate, Energy and Utilities, 2019).

Despite the fact that the final scores of Finland are significantly lower than those of the first two countries, Finland firmly holds third position in all weighting scenarios. The indicators selected to reflect social dimension are among the best in Northern European countries, where SO1 and SO2 indicators are the highest among all selected countries. Despite the significant achievements in social aspects, the environmental indicators require improvements. For example, RES in heating energy mix counts only 44 percent, incineration of non-RES wastes counts 3 percent. While, solid fossil fuels meet a third of heating needs, of which half is hard coal and 15 percent counts natural gas. The energy mix of the country has a clear relationship with the amount of GHG emissions, which are for about 3 ton/per capita. District heating counted 46 percent in 2018; this indicator has a tendency to increase annually by 1–2 percent. District heating is popular in residential blocks, where the market share counts for about 90 percent. While in detached houses the share of district heating reached only 8 percent in 2018. The most common type of heating in detached houses is electricity. As do other climate-conscious countries, Finland seeks also to create a carbon-neutral energy system. Therefore, high-level taxes for fossil fuels were set and support mechanisms for generation of RES-based electricity were determined (Finland Ministry of Economic Affairs and Employment, 2019).

Estonia took the fourth position in the S1, S2 and S3 scenarios, with 56 percent share of RES in the heating fuel mix. According to the National Energy and Climate Plan 2030, the share of RES will be 63 percent by the end of 2030. In 2018, fossil fuels counted almost 40 percent. Despite that, the electricity generation is also based on solid fossil fuels (more than 76 percent in 2018). It should be highlighted that Estonia stands out from the other Baltic states with high achievements in the social dimension, where indicators are close to the values of Denmark and the UK. Despite a significant share of renewables in the heating mix, the district heating in the country is notable for the very high carbon content of oil shale. Together with the electricity sector, which relies on fossil fuels, the electricity and heating sector has very high values of GHG emissions per capita. Estonian district heating counted for about 60 percent. According to the National Energy and Climate Plan 2030, RES will count for about 80 percent of district heating in Estonia (Estonia's 2030 National Energy and Climate Plan, 2019). The biggest potential for development are heat pumps.

The statistical data shows that in the UK half of the energy consumed is for heating. Almost all heat energy was generated from fossil fuels in 2018, where natural gas accounted more than 90 percent and RES accounted for less than 10 percent.

As a consequence, heat was the main source of GHG emissions in the country, which contributed one third of national GHG emissions. The UK government also seeks to make the country carbon neutral by 2050. In order to encourage the RES usage in heating, the UK government introduced the Renewable Heat financial incentive in 2011, which funded biomass heat technologies. Jeswani et al. (2019) applied LCA and life cycle cost (LCC) for the analysis of biomass heat sustainability in the UK, from economic and environmental and point of view. According to the results, heat from solid biomass can reduce GHG emissions, eutrophication, and acidification, but the impacts on human health are worse than from natural gas. Also, the biomass sources available in the country are too small to meet all heating needs. The UK took last place in the ranking in the environmental scenario (S4). The district heating network is very small in the country. The government seeks to reach a target for 17 percent for district heating by 2030.

Lithuania took the last place in the ranking according to the three first scenarios. Despite quite good achievements in environmental dimension, gas prices were high, gas accounted for about one third of the heating needs. Also, the indicators reflecting the social dimension were very low: 28 percent of people did not heat their homes enough, and 9 percent had problems paying their energy bills on time. In the environmental scenario, the country rose from the bottom to the fourth position. Two thirds of energy in the heating mix were generated from RES in 2018. Also, the level of GHG emissions was low and counted only 0.36 ton/per capita. The district heating accounted for 57 percent, and 70 percent of heat was generated from RES and municipal waste.

Latvia took the penultimate position in the three first scenarios, having, as does Lithuania, very low indicators reflecting the social dimension. In Latvia, 24 percent of the people lived in buildings with leakages and 12 percent were unable to pay their utility bills on time. In the case of the environmental scenario (S4), Latvia rose by one place, leaving Estonia and the UK behind. Despite that, the increase in the ranking was not because of the good environmental indicators, but because of worsened environmental indicators of Estonia and the UK. RES accounted for 44 percent in the heating energy mix in Latvia in 2018. But more than a half of this heat comes from fossil fuels usage, where natural gas accounted for 53 percent.

The decarbonization of the heating sector is a very important part toward the global low-carbon energy transition. Therefore, energy policy should be focused on innovative technologies adaptation and development of the necessary legal basis. Also, it is very important to monitor the effectiveness of all implemented measures and their impact on economic, environmental, and social aspects.

3.2.4 FRAMEWORK TO MONITOR TRANSPORT SECTOR SUSTAINABILITY

This subsection presents a framework for the road transport sustainability assessment. The framework was created to measure sustainability in the EU countries and has been applied for the analysis of achievements made in the last decade (2010–2020) (Siksnelyte-Butkiene and Streimikiene (2022). The proposed framework accounts for the main indicators reflecting road transport sustainability issues and is a suitable tool to measure sustainability of the sector or monitor the progress achieved. It was followed by the provision that the developed framework should be easily applied in future studies. Therefore, much attention was paid to the indicators availability and the research instrument itself.

3.2.4.1   Energy Policy Context

The most critical sustainability issues that need to be addressed in the transport sector are linked to climate change mitigation. Therefore, the transport sector and its decarbonization play an important role in decarbonization of the whole EU economy (Brostow and Nellthorp, 2000). It is recognized that the EU is a flagship for low-carbon energy transition around the world. In 2019, the European Parliament issued a call for the European Commission to take into account climate emergency and to propose policy measures to fight against global warming and to address 1.5°C target (Papadis and Tsatsaronis, 2020). In 2021, the European Parliament adopted the EU Climate Law, where the new GHG reduction target of 55 percent by 2030 was introduced, and achievement of climate neutrality by 2050 in the entire region was declared. Such initiatives confirm the EU leadership to combat climate change. In pursuit of such ambitious goals, the European Commission established the European Green Deal, which can be recognized as the fundamental document to decarbonize the economy by 2050. To implement its ambitious targets regarding climate change, the EU has initiated legislation and regulations revision for sectors that have a direct impact on GHG emissions under the Fit for 55 package. The transport sector is the only one in the EU economy in which GHG emissions have increased more than 25 percent since 1990. The transport sector contributes to more than one fifth of total GHG emissions in the EU and is very important for climate change policy (European Commission et al., 2021).

The Fit for 55 package established in 2021 by the European Commission includes specific legal regulations, allowing implementation of the Green Deal objectives. The existing legislation on GHG emissions reduction in the energy sector was revised while preparing the package. Also, the European Commission adopted amendments of Renewable Energy Directive II, where the GHG reduction target for 2030 was proposed instead of the share of RES in transport. The new package consists of 13 related amended legislations and 6 new recommended laws related to climate change and energy.

The Renewable Energy Directive (2009/28/EC) (European Parliament and Council of the European Union, 2009) adopted by the European Commission in 2009 set the RES objectives for EU countries by 2020 to achieve 10 percent the share of RES in transport. The Renewable Energy Directive provides also sustainability criteria for renewable energy used in transport and based on these requirements only biofuels can comply with criteria set since 2011 (Ebadian et al., 2020). The new Renewable Energy Directive (2018/2001) was issued by the European Commission in 2018 (European Parliament and Council of the European Union, 2018b). The new directive strengthened the requirements for bioenergy and set the 14 percent RES target for transport until 2030.

The largest road transport polluters are cars and vans, which emit more than 15 percent of total GHG emissions (Qin-Lei et al, 2022). As a result, the European Commission declared a proposal to attain zero emissions from cars and vans by 2035 in the region. Also, the interim targets were set, which is reduction of emissions by 55 percent for cars and 50 percent for vans by 2030. However, it should be admitted that electrification of road transport will play the most important role in transport decarbonization (Di Felice et al., 2021; Mehlig et al., 2021). Also, it is very important to mention that despite the fact that electric and hydrogen vehicles are carbon-free, the electricity used to charge electric vehicles or produce hydrogen can be generated from fossil fuels (Blanco et al., 2018; 2019). Electricity generation still is dependent on fossil fuels. Therefore, it is crucial to decarbonize the electricity sector in order to reach effectiveness of transport electrification as an emission reduction instrument. Otherwise, road transport electrification will raise the overall GHG emissions level (de Tena and Pregger, 2018; Zhang and Fujimori, 2020).

The main policy measures encouraging decarbonization of the transport sector are based on financial initiatives, regulation, the development of required infrastructure, dissemination of information, and actions regarding the increase of people's awareness (Forrest et al., 2016; Thiel et al., 2016; Franzo and Nasca, 2021). All these measures are being used in the EU countries with the aim to overcome the most influencing barriers for the smooth development of carbon-free vehicles such as affordability, convenience, and awareness (Perdiguero and Jimenez, 2012; Troitino, 2015; De Gennaro et al., 2016; Tattini et al., 2018). The financial incentives enable boosting the initial uptake of electric vehicles and getting scale economy in vehicles and battery production. Such measures as subsidies for purchasing a vehicle and rebates on registration tax are very popular in many countries because they reduce the price gap between electric and traditional fossil fuels based vehicles (European Commission, 2013; Liu et al., 2021). The revision of CO2 emissions standards for traditional fossil fuels based vehicles are another effective measure to encourage penetration of electric vehicles into the market (de Blas et al., 2020). But the main measure to scale up electric vehicles is the development of required infrastructure, which would be convenient and affordable for the users. The direct investments for the publicly accessible chargers' installation is among the most popular support measures in many countries (Zhang and Fujimori, 2020). In some countries, there is a requirement for the new constructed buildings to include charging stations for electric vehicles. The differentiated circulation fees for carbon-free and traditional vehicles or establishment free of charge zero carbon parking areas are also among the most popular measures to stimulate road transport decarbonization.

3.2.4.2   Indicators for Comparative Assessment of Road Transport Sustainability

As with all presented illustrative examples of multi-criteria application for sustainability assessment, the presented framework follows recommendations of the Bellagio STAMP guidelines, where the importance of the easy application in the future is indicated. In order to ensure that, all the indicators were selected in consideration of their importance to the main purpose of the evaluation, public availability, and other recommendations. The developed framework can be applied not only for countries' comparison, but also for the analysis of selected countries achievements year by year. But the full view of the entire region and the effect of the policy measures used can be seen only when all EU member states are under analysis.

Six indicators have been selected for the evaluation of road transport sustainability and countries' ranking. These indicators can be arranged into three groups, which reflect environmental (the share of RES in transport sector; the share of passenger diesel cars), mobility (the number of passenger cars per 1,000 inhabitants and the share of buses and trains in inland passenger transport); and health aspects (amount of GHG emissions from fuel combustion in road transport per person; rate of road traffic deaths).

The share of renewable energy and reduction of polluting vehicles are among the EU's transport and climate change policy objectives (Ebadian et al., 2020; Qin-Lei et al., 2022). Therefore, such indicators as the share of RES and the share of passenger diesel vehicles are included in the assessment. For the sustainable city development and welfare of people, a well-developed public transport system is necessary. Also, the created mobility options significantly affect air quality and people's health (Schulte-Fischedick et al., 2021). The mobility situation in a specific country is reflected by the level of development of the public transport system and the use of private cars in inland passenger transport. The GHG emissions reduction and the decrease of road traffic deaths are also among the EU objectives (European Commission et al., 2021; European Commission et al., 2021) and they reflect road transport to the people's health.

The selected road transport indicators for sustainability assessment of EU countries in 2010 and 2020 are presented in Table 3.25 and Table 3.26.

The latest statistical data analysis showed that Sweden stands out from the other countries with the impressive results of the portion of RES in the transport sector. Almost one third (31.9%) of the energy consumed in transport is carbon-free. In most other countries, the share of RES varies between 9–12 percent. However, in countries such as Croatia, Cyprus, Greece, Latvia, Lithuania, and Poland, the share of RES is really small and counts for only 5–7 percent. These results indicate that urgent measures should be implemented in these lagging countries in order to achieve EU transport policy targets.

Europe seeks to reduce the number of polluting diesel vehicles to a minimum as soon as possible. However, the popularity of diesel vehicles varies among countries: in some there is a tendency to rise despite the tightening regulation, legislation, and various fiscal measures. In some countries, the share of passenger diesel cars counts for more than a half of all passenger cars: these are: Lithuania (68%), Latvia (63%), Bulgaria (59%), Ireland (59%), France (58%), Portugal (57%), Austria (55%), Croatia (55%), Spain (54%), Luxembourg (53%,) and Slovenia (50%). In some of these countries, parking is also quite old (e.g., Bulgaria, Latvia, Lithuania). The share of passenger diesel cars has stopped growing in Latvia and Lithuania over the past few years, but clear downward trends are not observed yet. The increase in popularity of buying diesel cars still is observed in Bulgaria, most of which are old and polluting cars from Western Europe.

A well-developed public transport infrastructure, which ensures the people's needs and smooth transportation services can significantly contribute to the state of the environment and ensure the quality of peoples' life. The notable differences of the share of buses and trains in inland passenger transport were observed among the EU member states. The smallest portion of public transport was observed in Lithuania,

TABLE 3.25
Road Transport Indicators in EU Countries in 2010

Countries	Renewable energy in transport sector, %	Passenger diesel cars, %	Share of buses and trains in inland passenger transport, %	Passenger cars per 1000 inhabitants*, number	GHG emissions from fuel combustion in road transport, tonnes/ person	Road traffic deaths, rate
Austria	10.7	55.08	20.4	530	2.611	6.6
Belgium	4.8	60.29	19.8	480	2.354	7.8
Bulgaria	1.5	36.40	20	353	1.024	10.5
Croatia	1.1	35.24	16.3	355	1.327	9.9
Cyprus	2.0	9.94	18.1	551	2.874	7.2
Czech Republic	5.2	26.82	26.5	429	1.558	7.7
Denmark	1.2	23.02	20.3	394	2.229	4.6
Estonia	0.4	24.81	16.4	416	1.568	5.9
Finland	4.4	19.29	15.1	535	2.202	5.1
France	6.6	61.91	14.5	487	1.957	6.2
Germany	6.4	26.63	14.0	527	1.811	4.5
Greece	1.9	1.25	18.4	469	1.752	11.3
Hungary	6.2	20.77	31.5	299	1.138	7.4
Ireland	2.5	27.07	17.4	424	2.409	4.6
Italy	4.9	37.82	18.3	619	1.786	6.9
Latvia	4.0	33.00	21.8	307	1.442	10.4
Lithuania	3.8	15.18	8.3	554	1.331	9.7
Luxembourg	2.1	63.73	16.5	659	12.837	6.3
Malta	0.4	27.94	18.5	581	1.226	3.1
The Netherlands	3.4	16.77	13.5	464	1.998	3.2
Poland	6.6	22.45	23.9	453	1.272	10.3
Portugal	5.5	44.27	10.9	444	1.731	8.9
Romania	1.4	30.74	22	214	0.651	11.7
Slovakia	5.3	7.50	22.2	310	1.206	6.9
Slovenia	3.1	34.62	13.2	518	2.569	6.7
Spain	5.0	51.77	17.7	475	1.820	5.2
Sweden	9.6	13.99	15.4	460	2.080	2.8

Source: Created by authors based on data from Eurostat (2022) and European Environment Agency (2019).

* the data for Denmark is not available in 2010, therefore the data for 2011 is included.

with only 5.8 percent in all inland passenger transport. In Lithuania, the services of public transport are not popular, and people prefer private cars not only for long trips, but also for short ones. Such preferences are due to many reasons, but the major role is convenience. Therefore, the low-usage of public transport services is determined by transport infrastructure planning, such as availability, affordability, travel time, comfort level, etc.

TABLE 3.26
Road Transport Indicators in EU Countries in 2020

Countries	Renewable energy in transport sector, %	Passenger diesel cars*, %	Share of buses and trains in inland passenger transport, %	Passenger cars per 1000 inhabitants, number	GHG emissions from fuel combustion in road transport, tonnes/ person	Road traffic deaths, rate
Austria	10.3	54.53	19.4	570	2.307	3.9
Belgium	11.0	48.12	13.5	510	1.801	4.3
Bulgaria	9.1	59.1	10.4	414	1.329	6.7
Croatia	6.6	54.47	11.1	433	1.386	5.9
Cyprus	7.4	21.38	12.4	645	2.156	5.4
Czech Republic	9.4	39.48	18.3	565	1.628	4.8
Denmark	9.7	30.38	12.7	466	1.933	2.7
Estonia	12.2	40.40	11.6	608	1.635	4.4
Finland	13.4	26.11	13.0	652	1.797	4.0
France	9.2	57.46	13.0	567	1.546	3.7
Germany	9.92	31.21	11.2	580	1.721	3.3
Greece	5.3	8.10	12.9	514	1.248	5.5
Hungary	11.6	31.68	21.2	403	1.264	4.7
Ireland	10.2	58.69	13.6	458	1.946	2.9
Italy	10.7	43.88	14.1	670	1.324	4.0
Latvia	6.7	63.34	12.3	390	1.583	7.3
Lithuania	5.5	67.72	5.8	560	2.116	6.2
Luxembourg	12.6	53.18	13.4	682	7.407	4.1
Malta	10.6	31.57	13.8	597	1.035	2.3
The Netherlands	12.6	12.54	9.9	503	1.459	3.0
Poland	6.6	31.51	12.4	664	1.638	6.6
Portugal	9.7	56.90	6.8	540	1.392	5.2
Romania	8.5	48.36	18.1	379	0.921	8.5
Slovakia	9.3	44.30	18.8	447	1.248	4.5
Slovenia	10.9	50.29	8.7	555	2.169	3.8
Spain	9.5	54.44	10.0	521	1.477	2.9
Sweden	31.9	35.39	16.0	476	1.365	2.0

Source: Created by authors based on data from Eurostat (2022), European Environment Agency (2019), The European Automobile Manufacturers' Association (2021).

* There is no available data from Bulgaria, Greece and Slovakia in 2020, therefore the data for 2019 is included for Greece and Slovakia, the data for 2017 is included for Bulgaria.

Also, the number of passenger cars significantly differs among the EU member states, from the 379 cars per 1,000 inhabitants in Romania to 682 for 1,000 inhabitants in Luxembourg. Such disparities are caused by many factors, including geographical aspects, affordability, distribution of population, tax policy, the development of public transport infrastructure, and so forth.

The GHG emissions from fuel combustion in road transport was 1.54 tonnes/person in the EU-28 in 2020. Although the amount of GHG emissions for one country inland habitant range from 0.92 in Romania to 7.41 in Luxembourg, mostly the amount of emissions is similar to the EU-28 average in many countries. However, some countries stand out with really high values of emissions: Luxembourg (7.4), Austria (2.3), Slovenia (2.2), Cyprus (2.2), Lithuania (2.1), Ireland (2), Denmark (1.9). International road traffic has a considerable influence on such results. Luxembourg stood out with the highest values of GHG emissions per person across the EU. Such results are influenced by the geographical location of the country, which is at the core of the main traffic axes for Western Europe (freight and passengers). In Luxembourg, domestic traffic is responsible for only 25 percent of the road transport fuels sold.

The rate of deaths in road traffic indicates the rate of persons killed in road accidents per the average number of people in the country. It necessary to highlight that this indicator differs between some countries by more than four times. For example, in Sweden the indicator value is 2, while in Romania it is 8.5. In four EU member states – Bulgaria, Romania, Latvia, Poland – the indicator still shows a very bad situation regarding road traffic fatalities, though the clear decrease trends since 2010 can be noticed in the other EU countries. The lowest number of road fatalities in the EU was in 2020 and counted for 18,800 fatalities. Such a decrease was impacted by the COVID-19 pandemic and lockdowns introduced (ITF, 2021). In 2021, the number of fatalities increased and reached 19,800. However, the rate is significantly lower than before the pandemic. The EU has a target for 2030 to reduce the number of road fatalities to 11,400 or less (European Commission, 2022b). Safety on roads is dependent on many aspects, such as: the level of public awareness; transport infrastructure compliance with the needs; regulatory system, and so forth.

The whole transport sector, and especially road transport, plays a significant role in terms of sustainable energy sector development (Ammermann et al., 2015) as the sector has direct impact on the level of GHG emissions, atmospheric pollution, people's well-being, and so forth. The increase of usage of public transport services and the reduction of passenger cars, especially diesel cars, allow for reduction of the burden of climate change and reducing GHG emissions levels from the transport sector, which affects the whole environment and peoples' health (de Blas et al., 2020). The other important stimulus for the sustainable transport sector is the portion of RES and fast penetration of electric and hybrid vehicles.

The analysis of road transport data showed that countries differ significantly in terms of sustainability of road transport across the EU. Therefore, it is necessary to implement transport policy taking into account sustainability issues and to monitor the countries' achievements in the whole region. Looking at the data in the last decades, many countries have achieved notable results regarding transport sector sustainability. However, this is not enough to fight against climate change, and much more effort should be given for the goals of energy transition.

3.2.4.3 MCDM Technique

For computation of results achieved and countries ranking in the ten years' period 2010–2020, the TOPSIS method was applied. The detailed presentation of TOPSIS

TABLE 3.27
Weighting Schemes for Sensitivity Analysis

	Environment		Mobility		Health	
Weighting schemes	RES in transport	Passenger diesel cars	Buses and trains in inland passenger transport	Share of passenger cars	GHG emissions in road transport	Road traffic deaths
S1 (Basic)	16.67 %	16.67 %	16.67 %	16.67 %	16.67 %	16.67 %
S2 (Mobility)	12.5 %	12.5 %	25 %	25 %	12.5 %	12.5 %
S3 (Health)	12.5 %	12.5 %	12.5 %	12.5 %	25 %	25 %
S4 (Environment)	25 %	25 %	12.5 %	12.5 %	12.5 %	12.5 %

Source: Created by authors.

techniques step by step can be found in the subsection 3.2.2. For the evaluation of improvements made during the period 2010–2020, a basic scenario was created, where equal weights for all indicators were assumed. In order to validate the results, sensitivity analysis was performed with the creation of three additional scenarios, which focus on mobility, health, and environmental issues. Created scenarios and weighting schemes are provided in Table 3.27.

3.2.4.4 Multi-Criteria Assessment Results

The evaluation results and countries' rankings are provided in Table 3.28. The performed assessment in two time ranges allow for estimating countries' achievements made during the decade. According to the results, the leading country in the EU regarding road transport sector sustainability is Sweden, with the score of relative distance to the ideal solution, 0.8415 in 2020, and 0.7901 in 2010. In 2020, the results of the countries ranked in the rest positions do not differ to a great extent. However, the results of Luxembourg are extremely low, with the score 0.2575 in the ranking in 2020 and 0.162 in 2010. Such a low score was influenced by a significant deviation of GHG emissions compared with other countries under assessment. The country is unique from the other EU countries, because Luxembourg is a center for international road traffic in Europe.

The results of calculations showed that the leading countries in terms of sustainable transport development in EU are Sweden and new EU member states such as Hungary, Slovakia, Poland, and the Czech Republic in 2010. The situation has changed in 2020 and the Netherlands, Malta, and Finland appeared among the best performing countries.

The high ranking of Slovakia, Hungary, Denmark, and the Czech Republic in terms of road transport sustainability in 2010 is mainly due to a large portion of public transport (leaders are Hungary, Czech Republic, Poland, Slovakia), quite a low share of passenger cars (leaders are Romania, Hungary, Latvia, Slovakia), low

TABLE 3.28

Road Transport Sustainability and Achievements Made in the EU Countries

Countries	2010				2020				Achievements, 2010–2020
	S_i^+	S_i^-	C_i	Rank	S_i^+	S_i^-	C_i	Rank	
Sweden	0.038	0.144	0.7901	1	0.024	0.130	0.8415	1	0
Hungary	0.041	0.144	0.7794	2	0.062	0.108	0.6337	2	0
Slovakia	0.044	0.143	0.7630	3	0.071	0.104	0.5933	6	−3
Poland	0.051	0.138	0.7287	4	0.084	0.092	0.5226	23	−19
Czech Republic	0.052	0.133	0.7192	5	0.072	0.098	0.5769	11	−6
Germany	0.054	0.132	0.7095	6	0.071	0.097	0.5791	10	−4
Austria	0.063	0.133	0.6777	7	0.074	0.090	0.5485	18	−11
Finland	0.061	0.126	0.6758	8	0.062	0.098	0.6124	5	3
The Netherlands	0.063	0.130	0.6721	9	0.062	0.107	0.6333	3	6
Latvia	0.065	0.130	0.6651	10	0.091	0.091	0.4982	25	−15
Italy	0.066	0.125	0.6552	11	0.071	0.100	0.5871	7	4
Spain	0.069	0.125	0.6455	12	0.077	0.098	0.5614	14	−2
Lithuania	0.073	0.131	0.6434	13	0.099	0.081	0.4496	26	−13
Ireland	0.069	0.121	0.6385	14	0.075	0.094	0.5562	15	−1
Portugal	0.071	0.125	0.6368	15	0.083	0.094	0.5325	20	−5
Greece	0.075	0.132	0.6364	16	0.081	0.105	0.5638	13	3
Romania	0.079	0.137	0.6336	17	0.084	0.104	0.5532	17	0
Denmark	0.072	0.124	0.6318	18	0.069	0.097	0.5837	9	9
France	0.074	0.125	0.6290	19	0.078	0.096	0.5534	16	3
Bulgaria	0.080	0.130	0.6195	20	0.084	0.095	0.5290	21	−1
Cyprus	0.073	0.119	0.6193	21	0.080	0.089	0.5277	22	−1
Estonia	0.079	0.128	0.6166	22	0.068	0.096	0.5837	8	14
Malta	0.083	0.132	0.6155	23	0.066	0.110	0.6250	4	19
Slovenia	0.075	0.115	0.6067	24	0.076	0.087	0.5332	19	5
Croatia	0.082	0.126	0.6049	25	0.086	0.094	0.5223	24	1
Belgium	0.078	0.118	0.6032	26	0.071	0.093	0.5675	12	14
Luxembourg	0.157	0.030	0.1620	27	0.119	0.041	0.2575	27	0

Source: Created by authors.

value of GHG emissions from fuel combustion in transport (leaders are Romania, Bulgaria, Hungary, Slovakia) and a low share of passenger diesel cars (leaders are Greece, Slovakia). At the same time, Sweden, Malta, and the Netherlands stand out with a really low road traffic fatalities rate. The biggest portion of RES in the transport sector was observed in Sweden in all periods under analysis and will probably maintain those position for some time in the future.

In 2020, the leading countries are Sweden, Hungary, the Netherlands, Malta, and Finland. The main reasons for the high ranking of these countries are the decrease of GHG emissions from fuel combustion in transport due to the increased portion of RES and the fast penetration of carbon-free cars. The adopted tight standards for road vehicles including financial measures such as subsidies for buying electric vehicles, annual tax exemptions on vehicles registration and operation, and the development of required transport infrastructure allowed to reach proper results. The low road fatalities rate in Sweden, Malta, Denmark, and the Netherlands was another significant contributor to high results of the leading countries in terms of sustainability. The high position of Hungary was determined by a large share of RES, low number of passenger cars, and a large portion of busses and trains in inland passenger transport.

The development and penetration of carbon-free technologies are critical for the future transport sector. Urgent actions are necessary to boost renewable energy in transport for such countries as Greece, Lithuania, Latvia, Poland, Croatia, and Cyprus. A good example is Sweden with impressive results regarding RES development in transport.

In summary, the countries perform better in terms of a high share of public transport and low number of passenger cars in East and Central Europe. While the Nordic countries mainly are leading because of their low road traffic deaths rate and lower portion of passenger diesel cars.

Last in the assessment during the investigated period remained Luxembourg, distinguished by the highest level of GHG emissions from fuel combustion in transport. Also, Luxembourg has the highest rate of passenger cars per inhabitant in the EU. Also, the country had the biggest share of diesel passenger cars in 2010 in the EU, and though the situation slightly improved in 2020, Luxembourg still remained among the leaders according the usage of diesel vehicles.

In order to validate the results for weighting, scenarios were created. The results comparison is provided in Table 3.29 and Table 3.30.

The utility scores coefficients correlation among different weighting schemes show a high correlation (Table 3.31 and Table 3.32). The lowest coefficient is 0.846 for the health and environment scenarios in 2010. Also, for the same scenarios the lowest coefficient is 0.773 in 2020. Despite that, the correlation is high and proves that despite the different weightings, the evaluation results can be considered as validated.

The conducted research results can be compared with other studies dealing with policy and development trends of the transport sector in the EU. The differences between the old and new EU member states have been mostly analyzed. For example, data envelopment analysis (DEA) was applied for the efficiency measurement of road and rail freight transport in EU member states by Baran and Gorecka (2019). According to the results, there are no significant inland transport efficiency disparities in old and new EU countries. Also, the authors did not identify statistical correlations between the economic condition of a country and its road transport efficiency. The measurement and analysis of road transport sustainability revealed that, in terms of road transport sustainability issues, some differences between old and new EU members can be found. These differences are regarding the investments in new technologies that reduce GHG emissions and encourage RES penetration in the transport sector. Stefaniec et al. (2021) applied DEA analysis for the analysis of old and new EU countries in terms of social sustainability. Such aspects as mobility, accessibility,

TABLE 3.29
Results Comparison of Different Weighting Schemes, 2010

	S1		S2		S3		S4	
Country	Utility	Rank	Utility	Rank	Utility	Rank	Utility	Rank
Sweden	0.7901	1	0.6849	5	0.8447	1	0.8145	1
Hungary	0.7794	2	0.7972	1	0.8276	2	0.7056	2
Slovakia	0.7630	3	0.7427	2	0.8248	3	0.6993	3
Poland	0.7287	4	0.7012	4	0.7669	8	0.6876	4
Czech Republic	0.7192	5	0.7098	3	0.7896	5	0.6324	6
Germany	0.7095	6	0.6158	10	0.8052	4	0.6596	5
Austria	0.6777	7	0.6321	8	0.7485	13	0.6302	7
Finland	0.6758	8	0.5980	16	0.7737	7	0.6081	8
The Netherlands	0.6721	9	0.6015	15	0.7821	6	0.5888	10
Latvia	0.6651	10	0.6630	6	0.7365	17	0.5639	13
Italy	0.6552	11	0.5875	17	0.7600	10	0.5650	12
Spain	0.6455	12	0.6027	13	0.7644	9	0.5275	17
Lithuania	0.6434	13	0.5501	25	0.7335	18	0.5904	9
Ireland	0.6385	14	0.6022	14	0.7499	12	0.5266	18
Portugal	0.6368	15	0.5665	24	0.7333	19	0.5535	14
Greece	0.6364	16	0.6037	12	0.7060	26	0.5690	11
Romania	0.6336	17	0.6444	7	0.7163	22	0.5108	19
Denmark	0.6318	18	0.6174	9	0.7483	14	0.5093	20
France	0.6290	19	0.5743	21	0.7441	15	0.5288	16
Bulgaria	0.6195	20	0.6120	11	0.7185	20	0.4895	23
Cyprus	0.6193	21	0.5716	23	0.7093	25	0.5434	15
Estonia	0.6166	22	0.5855	18	0.7439	16	0.4902	22
Malta	0.6155	23	0.5738	22	0.7547	11	0.4828	24
Slovenia	0.6067	24	0.5423	26	0.7176	21	0.5049	21
Croatia	0.6049	25	0.5831	19	0.7127	23	0.4761	26
Belgium	0.6032	26	0.5788	20	0.7119	24	0.4826	25
Luxembourg	0.1620	27	0.1847	27	0.1555	27	0.1555	27

Source: Created by authors.

health, safety, employment, and equity in regional transportation were analyzed. The results showed that the new EU countries perform better in terms of social sustainability. These results are associated with a higher share of public transport and lower rate of motorization compared to the old EU countries. The results of current research also revealed that these aspects are specific for East and Central European countries. Therefore, the new EU member states perform better in terms of this aspect.

TABLE 3.30
Results Comparison of Different Weighting Schemes, 2020

Country	S1 Utility	S1 Rank	S2 Utility	S2 Rank	S3 Utility	S3 Rank	S4 Utility	S4 Rank
Sweden	0.8415	1	0.8029	1	0.8891	1	0.8193	1
Hungary	0.6337	2	0.6724	2	0.7372	4	0.5003	4
The Netherlands	0.6333	3	0.5838	7	0.7544	3	0.5382	2
Malta	0.6250	4	0.6023	4	0.7615	2	0.4898	5
Finland	0.6124	5	0.5743	9	0.7231	5	0.5125	3
Slovakia	0.5933	6	0.6227	3	0.7174	7	0.4414	11
Italy	0.5871	7	0.5638	11	0.7211	6	0.4486	10
Estonia	0.5837	8	0.5462	15	0.7075	11	0.4623	6
Denmark	0.5837	9	0.5705	10	0.7121	9	0.4551	8
Germany	0.5791	10	0.5436	16	0.7126	8	0.4544	9
Czech Republic	0.5769	11	0.5962	5	0.6950	13	0.4360	12
Belgium	0.5675	12	0.5591	12	0.6935	15	0.4289	13
Greece	0.5638	13	0.5531	14	0.6804	16	0.4555	7
Spain	0.5614	14	0.5277	18	0.7104	10	0.4125	16
Ireland	0.5562	15	0.5553	13	0.6939	14	0.4044	18
France	0.5534	16	0.5387	17	0.6975	12	0.4002	21
Romania	0.5532	17	0.5846	6	0.6367	23	0.4185	15
Austria	0.5485	18	0.5800	8	0.6640	19	0.4023	20
Slovenia	0.5332	19	0.4875	24	0.6656	18	0.4055	17
Portugal	0.5325	20	0.4875	24	0.6706	17	0.3961	22
Bulgaria	0.5290	21	0.5132	19	0.6472	21	0.3882	23
Cyprus	0.5277	22	0.5041	21	0.6379	22	0.4236	14
Poland	0.5226	23	0.4998	23	0.6358	24	0.4029	19
Croatia	0.5223	24	0.5110	20	0.6532	20	0.3736	24
Latvia	0.4982	25	0.5021	22	0.6134	25	0.3525	25
Lithuania	0.4496	26	0.4081	26	0.5847	26	0.3126	26
Luxembourg	0.2575	27	0.2887	27	0.2402	27	0.2620	27

Source: Created by authors.

TABLE 3.31
Utility Scores Correlation Coefficients, 2010

	S1	S2	S3
S1	1		
S2	0.951	1	
S3	0.968	0.897	1
S4	0.942	0.860	0.846

Source: Created by authors.

TABLE 3.32
Utility Scores Correlation Coefficients, 2020

	S1	S2	S3
S1	1		
S2	0.954	1	
S3	0.951	0.876	1
S4	0.924	0.872	0.773

Source: Created by authors.

APPENDIX 3.1 ABBREVIATIONS OF THE EU MEMBER STATES

AT – Austria
BE – Belgium
BG – Bulgaria
CY – Cyprus
CZ – Czech Republic
DE – Germany
DK – Denmark
EE – Estonia
ES – Spain
FI – Finland
FR – France
GR – Greece
HR – Croatia
HU – Hungary
IE – Ireland
IT – Italy
LT – Lithuania
LU – Luxembourg
LV – Latvia
MT – Malta
NL – The Netherlands
PO – Poland
PT – Portugal
RO – Romania
SE – Sweden
SI – Slovenia
SK – Slovakia

REFERENCES

Afsordegan, A., Sánchez, M., Agell, N., Zahedi, S., Cremades, L.V. Decision making under uncertainty using a qualitative TOPSIS method for selecting sustainable energy alternatives. *Int J Environ Sci Technol*, 2016, 13(6), 1419–32.

Alola, A.A., Saint Akadiri, S., Akadiri, A.C., Alola, U.V., Fatigun, A.S. Cooling and heating degree days in the US: The role of macroeconomic variables and its impact on environmental sustainability. *Sci Total Environ*, 2019, 695, 133832.

Ammermann, H., Ruf, Y., Lange, S., Fundulea, D., Martin, A. Fuel Cell Electric Buses – Potential for Sustainable Public Transport in Europe. *Roland Berger*, 2015. Retrieved August 18, 2022, from www.fch.europa.eu/sites/default/files/150909_FINAL_Bus_Study_Report_OUT_0.PDF

Ayodele, T.R., Ogunjuyigbe, A.S.O. Increasing household solar energy penetration through load partitioning based on quality of life: The case study of Nigeria. *Sustain Cities Soc*, 2015, 18, 21–31.

Baran, J., Gorecka, A.K. Economic and environmental aspects of inland transport in EU countries. *Econ Res-Ekon Istraz*, 2019, 32(1), 1037–1059.

Bataille, C., Waisman, H., Colombier, M., Segafredo, L., Williams, J. The Deep Decarbonization Pathways Project (DDPP): Insights and emerging issues. *Clim Policy*, 2016. 16, S1–S6. http://dx.doi.org/10.1080/14693062.2016.1179620

Bellos, E., Tzivanidis, C. Energetic and financial sustainability of solar assisted heat pump heating systems in Europe. *Sustain Cities Soc*, 2017, 33, 70–84.

Bhardwaj, A., Joshi, M., Khosla, R., Dubash, N.K. More priorities, more problems? Decision-making with multiple energy, development and climate objectives. *Energy Res Soc Sci*, 2019, 49, 143–57.

Bhattacharyya, S.C. Review of alternative methodologies for analysing off-grid electricity supply. *Renew Sustain Energy Rev*, 2012, 16, 677–694.

Bhowmik, C., Bhowmik, S., Ray, A., Pandey, K.M. Optimal green energy planning for sustainable development: A review. *Renew Sustain Energy Rev*, 2017, 71, 796–813.

Bilgili F. The impact of biomass consumption on CO_2 emissions: Cointegration analyses with regime shifts. *Renew Sust Energ Rev*, 2012, 16(7), 5349–54.

Blanco, H., Gómez Vilchez, J.J., Nijs, W., Thiel, C., Faaij, A. Soft-linking of a behavioral model for transport with energy system cost optimization applied to hydrogen in EU. *Renew Sust Energ Rev*, 2019, 115, 109349.

Blanco, H., Nijs, W., Ruf, J., Faaij, A. Potential for hydrogen and Power-to-Liquid in a low-carbon EU energy system using cost optimization. *Appl Energy*, 2018, 232, 617–39.

Brauers, W.K.M., Zavadskas, E.K. MULTIMOORA Optimization Used to Decide on a Bank Loan to Buy Property, *Technol Econ Dev Econ*, 2011, 17(1): 174–88.

Brauers, W.K.M., Zavadskas, E.K. Project Management by MULTIMOORA as an Instrument for Transition Economies. *Technol Econ Dev Econ*, 2010, 16(1): 5–24.

Bristow, A.L., Nellthorp, J. Transport project appraisal in the European Union, *Trans Policy*, 2000, 7(1), 51–60.

Brown, M.A., Sovacool, B.K. Developing an "Energy Sustainability Index" to Evaluate American Energy Policy, Working Paper 18, 2007, School of Public Policy Georgia Institute of Technology.

Budzianowski, W.M. Negative carbon intensity of renewable energy technologies involving biomass or carbon dioxide as inputs. *Renew Sust Energ Rev*, 2012, 16(9), 6507–21.

Burgherr, P., Hirschberg, S., Brukmajster, D., and Hampel, J. Survey of criteria and indicators. New Energy Externalities Developments for Sustainability (NEEDS), Research Stream RS 2b: Energy technology roadmap and stakeholders perspective. Project cofounded by the European Commission within the Sixth Framework Programme (2002–2006), 2005. Paul Scherrer Institut, Villigen PSI (Switzerland).

Casquero-Modrego, N., Goni-Modrego, M. Energy retrofit of an existing affordable building envelope in Spain, case study. *Sustain Cities Soc*, 2019, 44, 395–405.

Castano-Rosa, R., Sherriff, G., Solis-Guzman, J., Marrero, M. The validity of the index of vulnerable homes: Evidence from consumers vulnerable to energy poverty in the UK. *Energy Sources B: Econ Plan Policy*, 2020, 15(2), 72–91.

Cavallaro, F. Multi-criteria decision aid to assess concentrated solar thermal technologies. *Renew Energy*, 2009, 34, 1678–85.

Chaton, C., Guillerminet, M.L. Competition and environmental policies in an electricity sector. *Energy Econ*, 2013, 36, 215–28.

Chen, Y.Z., Wang, J., Lund, P.D. Sustainability evaluation and sensitivity analysis of district heating systems coupled to geothermal and solar resources. *Energy Conv Manag*, 2020, 220, 113084.

Claudia, R.M., Martinez, M., Pena, R. Scenarios for a hierarchical assessment of the global sustainability of electric power plants in Mexico. *Renew Sustain Energy Rev*, 2014, 33, 154–60.

Connolly, D. Heat Roadmap Europe: Quantitative comparison between the electricity, heating, and cooling sectors for different European countries. *Energy*, 2017, 139, 580–93.

Corsatea, T.D., Giaccaria, S. Market regulation and environmental productivity changes in the electricity and gas sector of 13 observed EU countries. *Energy*, 2018, 164, 1286–97.

Cucchiella, F., D'Adamo, I., Gastaldi, M., Koh, S.C., Lenny, R.P. A comparison of environmental and energetic performance of European countries: A sustainability index. *Renew Sustain Energy Rev*, 2017, 78, 401–13.

Danish Ministry of Climate, Energy and Utilities. Denmark's Integrated National Energy and Climate Plan. 2019, 184 p.

De Blas, I., Mediavilla, M., Capellán-Pérez, I., Duce, C. The limits of transport decarbonization under the current growth paradigm. *Energy Strategy Rev*, 2020, 32, 100543.

De Gennaro, M., Paffumi, E., Martini. G. Big Data for Supporting Low-Carbon Road Transport Policies in Europe: Applications, Challenges and Opportunities. *Big Data Res*, 2016, 6, 11–25.

De Rubeis, T., Falasca, S., Curci, G., Paoletti, D., Ambrosini, D. Sensitivity of heating performance of an energy self-sufficient building to climate zone, climate change and HVAC system solutions. *Sustain Cities Soc*, 2020, 61, 102300.

De Tena, D.L., Pregger, T. Impact of electric vehicles on a future renewable energy-based power system in Europe with a focus on Germany. *Int J Energy Res*, 2018, 42(8), 2670–85.

Del Río, P., Resch, G., Ortner, A., Liebmann, L., Busch, S., Panzer, C.A techno-economic analysis of EU renewable electricity policy pathways in 2030. *Energy Policy*, 2017, 104, 484–93.

Di Felice, L.J., Renner, A., Giampietro, M. Why should the EU implement electric vehicles? Viewing the relationship between evidence and dominant policy solutions through the lens of complexity. *Environ Sci Policy*, 2021, 123, 1–10.

Dombi, M., Kuti, I., Balogh, P. Sustainability assessment of renewable power and heat generation technologies. *Energy Policy*, 2014, 67, 264–71.

Dziugaite-Tumeniene, R., Motuziene, V., Siupsinskas, G., Ciuprinskas, K., Rogoza, A. Integrated assessment of energy supply system of an energy-efficient house. *Energy and Build*, 2017, 138, 443–54.

Ebadian, M., van Dyk, S., McMillan, J.D., Saddler, J. Biofuels policies that have encouraged their production and use: An international perspective. *Energy Policy*, 2020, 147, 111906.

Ekholm, T., Karvosenoja, N., Tissari, J., Sokka, L., Kupiainen, K., Sippula, O., Savolahti, M., Jokiniemi, J., Savolainen, I. A multi-criteria analysis of climate, health and acidification impacts due to greenhouse gases and air pollution: The case of household-level heating technologies. *Energy Policy*, 2014, 74, 499–509.

Energy Community. Initiatives and infrastructure. Available online: www.energy-community.org/regionalinitiatives/NECP.html, 29 May 2022).

Eskeland, G.S., Rive, N.A., Mideksa, T.K. Europe's climate goals and the electricity sector. *Energ. Policy* 2012, 41, 200–211.

Estonia's 2030 National Energy and Climate Plan (NECP 2030). Estonia's Communication to the European Commission under Article 3(1) of Regulation (EU) No 2012/2018. 2019, 195 p. Available https://energy.ec.europa.eu/system/files/2022-08/ee_final_necp_main_en.pdf

European Commission, Clean Energy For All Europeans. Communication from the Commission to the European Parliament, the Council, the European Economic and Social Committee, the Committee of the Regions and the European Investment Bank. *Brussels, 30.11.2016 COM*. 2016a, 860. https://ec.europa.eu/energy/sites/ener/files/documents/com_860_final.pdf

European Commission, Directorate-General for Climate Action, Directorate-General for Energy, Directorate-General for Mobility and Transport, De Vita, A., Capros, P., Paroussos, L., et al. EU reference scenario 2020: Energy, transport and GHG emissions: Trends to 2050. *Publications Office*, 2021, 184 p. https://data.europa.eu/doi/10.2833/35750 (accessed on 10 September 2022).

European Commission, European Energy Security Strategy. Communication from the Commission to the European Parliament and the Council. Brussels, 28.5.2014 COM (2014) 330 final, 2014a. http://eur-lex.europa.eu/legal-content/EN/TXT/PDF/?uri=CELEX:52014DC0330&from=EN

European Commission, European Union Statistics on Income and Living Conditions (EU-SILC). *Eurostat, Brussels*, 2014c. Available at: https://ec.europa.eu/eurostat/web/income-and-living-conditions/data/database

European Commission, Proposal for a Directive of the European Parliament and of the Council amending Directive 2012/27/EU on energy efficiency. Brussels, 30.11.2016, COM(2016) 761 final 2016/0376 (COD), 2016b.

European Commission. A policy framework for climate and energy in the period from 2020 to 2030. Brussels, COM (2014) 15 final.

European Commission. An EU Strategy on Heating and Cooling. 2016c, Brussels, COM (2016) 51 final, 2014b.

European Commission. Energy 2020, A strategy for competitive, sustainable and secure energy. Communication from the Commission to the European Parliament, the Council, the European Economic and Social Committee and the Committee of the Regions. Brussels, 10.11.2010, COM (2010) 639 final, 2010a. http://eur-lex.europa.eu/legal-content/EN/TXT/PDF/?uri=CELEX:52010DC0639&from=EN

European Commission. Energy Roadmap 2050. 2011, Brussels, COM (2011) 885 final.

European Commission. Energy Union Package. A Framework Strategy for a Resilient Energy Union with a Forward Looking Climate Change Policy, Brussels, 2015, COM (2015) 80 final.

European Commission. Indicators for monitoring progress towards Energy Union objectives. 2022b. Available online: https://ec.europa.eu/energy/data-analysis/energy-union-ind icators/scoreboard_en?dimension=Energy+security%2C+solidarity+and+trust, 10 June 2022).

European Commission. Mobility and Transport. Road safety statistics: What is behind the figures? https://transport.ec.europa.eu/2021-road-safety-statistics-what-behind-figures_en (accessed on 22 September 2022), 2021.

European Commission. REPowerEU: Joint European Action for More Affordable, Secure and Sustainable Energy. Communication from the Commission to the European Parliament, the European Council, the Council, the European Economic and Social Committee and the Committee of the Regions. 2022a, Strasbourg, 8.3.2022, COM(2022) 108 final. Available online: (https://eur-lex.europa.eu/legal-content/EN/TXT/?uri=celex:5202 2DC0108, 26 May 2022).

European Commission. The JRC-EU-TIMES model: Assessing the long term role of the SET plan energy technologies. LU: Publications Office, 2013. Retrieved April 9, 2022, from https://data.europa.eu/doi/10.2790/97799

European Environment Agency. Data and maps. Dieselisation (share of diesel cars in the total passenger car fleet) in Europe. 2019. Available at: www.eea.europa.eu/ data-and-maps/daviz/dieselisation-of-diesel-cars-in-4#tab-chart_1 (accessed on 15 September 2022).

European Parliament; Council of the European Union. Directive (EU) 2018/2002 of the European Parliament and of the Council of 11 December 2018 amending Directive 2012/27/EU on energy efficiency. *Off J Eur Union*, 2018a, L 328/210.

European Parliament; Council of the European Union. DIRECTIVE (EU) 2018/2001 OF THE EUROPEAN PARLIAMENT AND OF THE COUNCIL of 11 December 2018 on the promotion of the use of energy from renewable sources. *Off J Eur Union*, 2018b, L 328/82, https://eur-lex.europa.eu/legal-content/EN/TXT/?uri=uriserv:OJ.L_ .2018.328.01.0082.01.ENG

European Parliament; Council of the European Union. DIRECTIVE 2009/28/EC OF THE EUROPEAN PARLIAMENT AND OF THE COUNCIL of 23 April 2009 on the promotion of the use of energy from renewable sources and amending and subsequently repealing Directives 2001/77/EC and 2003/30/EC. *Off J Eur Union*, 2009, 52, https://eur-lex.europa.eu/legal-content/EN/TXT/?uri= OJ:L:2009:140:TOC

European Parliament; Council of the European Union. Directive 2012/27/EU of the European Parliament and of the Council of 25 October 2012 on energy efficiency, amending Directives 2009/125/EC and 2010/30/EU and repealing Directives 2004/8/EC and 2006/ 32/EC. *Off J Eur Union*, 2012, L 315/1.

Eurostat. Database. 2022, Available at: https://ec.europa.eu/eurostat/data/database (accessed on 15 September 2022).

Finland Ministry of Economic Affairs and Employment. *Finland's Integrated Energy and Climate Plan*. 2019, 183 p. Helsinki, ISBN PDF: 978-952-327-478-5.

Flues, F., Löschel, A., Lutz, B.J., Schenker, O. Designing an EU energy and climate policy portfolio for 2030: Implications of overlapping regulation under different levels of electricity demand. *Energy Policy* 2014, 75, 91–99.

Forrest, K.E., Tarroja, B., Zhang, L., Shaffer, B., Samuelsen, S. Charging a renewable future: The impact of electric vehicle charging intelligence on energy storage requirements to meet renewable portfolio standards. *J Power Sources*, 2016, 336, 63–74.

Frank, L., Jacob, K., Quitzow, R. Transforming or tinkering at the margins? Assessing policy strategies for heating decarbonisation in Germany and the United Kingdom. *Energy Res Soc Sci*, 2020, 67, 101513.

Franzo, S., Nasca, A. The environmental impact of electric vehicles: A novel life cycle-based evaluation framework and its applications to multi-country scenarios. *J Clean Prod*, 2021, 315, 128005.

Goodbody, C., Walsh, E., McDonnell, K.P., Owende, P. Regional integration of renewable energy systems in Ireland – the role of hybrid energy systems for small communities. *Int J Electr Power Energy Syst*, 2013, 44(1), 713–20.

Gouveia, J.P., Palma, P., Simoes, S.G. Energy poverty vulnerability index: A multidimensional tool to identify hotspots for local action. *Energy Rep*, 2019, 5, 187–201.

Greening, L.A., Bernow, S. Design of coordinated energy and environmental policies: Use of multi-criteria decision-making. *Energy Policy*, 2004, 32, 721–35.

Harmsen, R., Wesselink, B., Eichhammer, W., Worrella, E. The unrecognized contribution of renewable energy to Europe's energy savings target. *Energy Policy*, 2011, 39(6), 3425–33.

Hehenberger-Risse, D., Straub, J., Niechoj, D., Lutzenberger, A. Sustainability Index to Assess the Environmental Impact of Heat Supply Systems. *Chem Eng Technol*, 2019, 42(9), 1923–7.

Herrera, I.; Rodrguez-Serrano, I.; Garrain, D.; Lechn, Y.; Oliveira, A. Sustainability assessment of a novel micro solar thermal: Biomass heat and power plant in Morocco. *J Ind Ecol*, 2020, 24, p. 1379–1392.

Hirschberg, S., Bauer, C., Burgherr, P., Dones, R., Schenler, W., Bachmann, T., Carrera, D.G. Environmental, economic and social criteria and indicators for sustainability assessment of energy technologies. New Energy Externalities Developments for Sustainability (NEEDS). Project cofounded by the European Commission within the Sixth Framework Programme (2002–2006). Paul Scherrer Institut, Villigen PSI (Switzerland). Project no: 502687, 2007.

Hobley, A. Will gas be gone in the United Kingdom (UK) by 2050? An impact assessment of urban heat decarbonisation and low emission vehicle uptake on future UK energy system scenarios. *Renew Energy*, 2019, 142, 695–705.

Hwang, C.L., Yoon, K. *Multiple Attributes Decision Making Methods and Applications*. Springer: Berlin, Heidelberg, 1981, 22–51.

IAEA. Energy indicators for sustainable development: guidelines and methodologies. International Atomic Energy Agency, United Nations Department of Economic and Social Affairs, International Energy Agency, Eurostat and European Environment Agency, 2005. www.unosd.org/content/documents/1237Pub1222_web%20EISD.pdf

Iddrisu I., Bhattacharyya, S.C. Sustainable Energy Development Index: A multi-dimensional indicator for measuring sustainable energy development. *Renew Sust Energ Rev*, 50, 2015, 513–30. https://doi.org/10.1016/j.rser.2015.05.032

IEA. *Technology Roadmap: Bioenergy for Heat and Power*. 2012, Paris, 68 p.

Ioannou, A., Angus, A., Brennan, F. Risk-based methods for sustainable energy system planning: A review. *Renew Sustain Energy Rev* 2017, 74, 602–15.

IRENA. Innovation Landscape for a Renewable-Powered Future: Solutions to Integrate Variable Renewables. *International Renewable Energy Agency, Abu Dhabi, United Arab Emirates*, 2019, ISBN 978-92-9260-111-9, 163.

ITF. Road Safety Annual Report 2021: *The Impact of Covid-19*, OECD Publishing, Paris, 2021, 67 p.

Jeswani, H.K., Whiting, A., Azapagic, A. Environmental and Economic Sustainability of Biomass Heat in the UK. *Energy Technol*, 2019, UNSP 1901044.

Jovanovic, M.P., Bakic, V.V., Vucicevic, B.S., Turanjanin, V.M. Analysis of Different Scenarios and Sustainability Measurement in the District Heating Sector in Serbia. *Therm Sci*, 2019, 23(3), 2085–96.

Kavvadias, K., Jimenez Navarro, J. and Thomassen, G., Decarbonising the EU heating sector: Integration of the power and heating sector, *EUR 29772 EN*, Publications Office of the European Union, Luxembourg, 2019, ISBN 978-92-76-08386-3.

Kirppu, H., Lahdelma, R., Salminen, P. Multicriteria evaluation of carbon-neutral heat-only production technologies for district heating. *Appl Therm Eng*, 2018, 130, 466–76.

Klein, S.J.W., Whalley, S. Comparing the sustainability of US electricity options through multi-criteria decision analysis. *Energy Policy*, 2015, 79, 127–49.

Knopf, B., Nahmmacher, P., Schmid, E. The European renewable energy target for 2030 – An impact assessment of the electricity sector. *Energy Policy*, 2015, 85, 50–60.

Kumar, A., Sah, B., Singh, A.R., Deng, Y., He, X., Kumar, P., Bansal, R.C. A review of multi criteria decision making (MCDM) towards sustainable renewable energy development. *Renew Sustain Energy Rev*, 2017, 69, 596–609.

Kurka, T., Blackwood, D. Selection of MCA methods to support decision making for renewable energy developments. *Renew Sustain Energy Rev*, 2013, 27, 225–33.

Kuznecova, I., Gedrovics, M., Kalnins, S.N., Gusca, J. Calculation framework of household sustainability index for heat generation. International Scientific Conference – Environmental and Climate Technologies, Conect 2016. *Energy Procedia*, 2017, 113, 476–81.

Kveselis, V., Dzenajaviciene, E.F., Masaitis, S. Analysis of energy development sustainability: The example of the Lithuanian district heating sector. *Energy Policy*, 2017, 100, 227–36.

Kyprianou, I., Serghides, D.K., Varo, A., Gouveia, J,P., Kopeva, D., Murauskaite, L. EP policies and measures in 5 EU countries: A comparative study. *Energy and Build*, 2019, 196, 46–60.

Lehmann, P., Sijm, J., Gawel, E., Strunz, S., Chewpreecha, U., Mercure, J.F., Pollitt, H. Addressing multiple externalities from electricity generation: A case for EU renewable energy policy beyond 2020? *Environ Econ Policy Stud*, 2019, 21(2), 255–83.

Ligus, M. Evaluation of Economic, Social and Environmental Effects of Low-Emission Energy Technologies Development in Poland: A Multi-Criteria Analysis with Application of a Fuzzy Analytic Hierarchy Process (FAHP). *Energies*, 2017, 10, 1550.

Liposcak, M., Afgan, N.H., Duic. N., Carvalho, M.G. Sustainability assessment of cogeneration sector development in Croatia. *Energy*, 2006, 31(13), 2276–84.

Liu, M., Zhang, Ch., Fan, Y., Jin, H. Cloud-Model-based MULTIMOORA Comprehensive Evaluation Method and Its Application in Statistics Discipline Assessment. *Transform Bus Econ*, 2019, 18, 3 (48), 229–45.

Liu, Z., Song, J., Kubal, J., Susarla, N., Knehr, K.W., Islam, E., Nelson, P., et al. Comparing total cost of ownership of battery electric vehicles and internal combustion engine vehicles. *Energy Policy*, 2021, 158, 112564.

Lockwood, M., Froggatt, A., Wright, G., Duttona, J. The implications of Brexit for the electricity sector in Great Britain: Trade offs between market integration and policy influence. *Energy Policy* 2017, 110, 137–43.

Lukanov, B.R., Krieger, E.M. Distributed solar and environmental justice: Exploring the demographic and socio-economic trends of residential PV adoption in California. *Energy Policy*, 2019, 134, UNSP 110935.

Maghmoumi, A., Marashi, F., Houshfar, E. Environmental and economic assessment of sustainable municipal solid waste management strategies in Iran. *Sustain Cities Soc*, 2020, 59, 102161.

Manolopoulosa, D., Kitsopoulosb, K., Kaldellisc, J.K. Bitzenis, A. The evolution of renewable energy sources in theelectricity sector of Greece. *Int Hydrogen Energ*, 2016, 41(29), 12659–71.

Mardani, A., Zavadskas, E.K., Khalifah, Z., Zakuan, N., Jusoh, A., Nor, K.M., Khoshnoudi, M. A review of multi-criteria decision-making applications to solve energy management problems: Two decades from 1995 to 2015. *Renew Sustain Energy Rev*, 2017, 71, 216–56.

Masini A., Menichetti E. Investment decisions in the renewable energy sector: an analysis of non-financial drivers. *Technol Forecast Soc Change*, 2013, 80(3), 510–24.

Mazhar, A.R., Liu, S., Shukla, A. A state of art review on the district heating systems. *Renew Sust Energ Rev*, 2018, 96, 420–39.

Mehlig, D., Woodward, H., Oxley, T., Holland, M., ApSimon, H. Electrification of Road Transport and the Impacts on Air Quality and Health in the UK. *Atmosphere*, 2021, 12(11), 1491.

Nami, H., Arabkoohsar, A., Anvari-Moghaddam, A. Thermodynamic and sustainability analysis of a municipal waste-driven combined cooling, heating and power (CCHP) plant. *Energy Conv Manag*, 2019, 201, 112158.

OECD, 2003. OECD environmental indicators: Development, measurement and use. Reference paper. www.oecd.org/env/indicators-modelling-outlooks/24993546.pdf

OECD, 2013. Environment at a Glance 2013: OECD Indicators, OECD Publishing. http://dx.doi.org/10.1787/9789264185715-en.

OECD/IEA, IRENA, 2017. Investment Needs for a Low-Carbon Energy System. *Perspectives for the Energy Transition*. Berlin. www.energiewende2017.com/wp-content/uploads/2017/03/Perspectives-for-the-Energy-Transition_WEB.pdf

OECD/NEA, 2002. Indicators of Sustainable Development in the Nuclear Energy Sector – A Preliminary Approach. NEA/NDC(2002)5.

Okushima, S. Understanding regional energy poverty in Japan: A direct measurement approach. *Energy and Build*, 2019, 193, 174–184.

Ozcan, E.C., Unlusoy, S., Eren, T. A combined goal programming–AHP approach supported with TOPSIS for maintenance strategy selection in hydroelectric power plants. *Renew Sust Energ Rev*, 2017, 78, 1410–23.

Pacesilaa, M., Burceaa, S.G., Colescab, S.E., 2016. Analysis of renewable energies in European Union. *Renew Sust Energ Rev*, April 2016, 56, 156–70. https://doi.org/10.1016/j.rser.2015.10.152

Papadis, E., Tsatsaronis, G. Challenges in the decarbonization of the energy sector. *Energy*, 2020, 205, 118025.

Perdiguero, J., Jimenez, J.L. Policy options for the promotion of electric vehicles: a review. *IRENA*, 2012. available online: www.ub.edu/irea/working_papers/2012/201208.pdf

Pereira, G.I., da Silva, P., Soule, D. Policy-adaptation for a smarter and more sustainable EU electricity distribution industry: a foresight analysis. *Environ Dev Sustain*, 2018, 20(Suppl 1), 231–267.

Pino-Mejias, R., Perez-Fargallo, A., Rubio-Bellido, C., Pulido-Arcas, J.A. Artificial neural networks and linear regression prediction models for social housing allocation: Fuel Poverty Potential Risk Index. *Energy*, 2018, 164, 627–41.

Pinto, G., Abdollahi, E., Capozzoli, A., Savoldi, L., Lahdelma, R. Optimization and Multicriteria Evaluation of Carbon-neutral Technologies for District Heating. *Energies*, 2019, 12(9), 1653.

Qin-Lei, J., Liu, H.Z., Yu, W.Q., He, X. The Impact of Public Transportation on Carbon Emissions: From the Perspective of Energy Consumption. *Sustainability*, 2022, 14(10), 6248.

Raghutla, C., Chittedi, K.R. Financial development, energy consumption, technology, urbanization, economic output and carbon emissions nexus in BRICS countries: an empirical analysis. *Manag Environ Qual*, Vol. 2020. https://doi.org/10.1108/MEQ-02-2020-0035

Recalde, M., Peralta, A., Oliveras, L., Tirado-Herrero, S., Borrell, C., Palencia, L., Gotsens, M., Artazcoz, L., Mari-Dell'Olmo, M. Structural energy poverty vulnerability and excess winter mortality in the European Union: Exploring the association between structural determinants and health. *Energy Policy*, 2019, 133, UNSP 110869.

Regeringen. *Sweden's Integrated National Energy and Climate Plan*. 2020, 199 p.

Ren, J., Lutzen, M. Selection of sustainable alternative energy source for shipping: Multi criteria decision making under incomplete information. *Renew Sustain Energy Rev*, 2017, 74, 1003–19.

Rodriguez-Alvarez, A., Orea, L., Jamasb, T. Fuel poverty and Well-Being: A consumer theory and stochastic frontier approach. *Energy Policy*, 2019, 131, 22–32.

RSC Project. The Low-Carbon Indicators Toolkit, 2016. www.rscproject.org/indicators/

Rutz, D., Worm, J., Doczekal, C., Kazagic, A., Duic, N., Markovska, N., Batas Bjelic, I.R., Sunko, R., Tresnjo, D., Merzic, A., Doracic, B., Gjorgievski, V., Janssen, R., Redzic, E., Zweiler, R., Puksec, T., Sunko, B., Rajakovic, N. Transition Towards a Sustainable Heating and Cooling Sector Case Study of Southeast European Countries. *Therm Sci*, 2019, 23(6), Part A, 3293–3306.

Saleem L., Ulfat I. A Multi Criteria Approach to Rank Renewable Energy Technologies for Domestic Sector Electricity Demand of Pakistan. *Mehran Univ Res J Eng Technol*, 2019, 38(2), 443–52.

Sangroya, D., Kabra, G., Joshi, Y., and Yadav, M. Green energy management in India for environmental benchmarking: From concept to practice. *Manag Environ Qual*, 2020, 31(5), 1329–49.

Sartori, S., Witjes, S., Campos, L.M.S. Sustainability performance for Brazilian electricity power industry: An assessment integrating social, economic and environmental issues. *Energy Policy*. 2017, 111, 41–51.

Schulte-Fischedick, M., Shan, Y.L., Hubacek, K. Implications of COVID-19 lockdowns on surface passenger mobility and related CO2 emission changes in Europe. *Appl Energy*, 2021, 300, 117396

Seddiki, M., Bennadji, A. Multi-criteria evaluation of renewable energy alternatives for electricity generation in a residential building. *Renew Sust Energ Rev*, 2019, 110, 101–117.

Shortall, R., Davidsdottir, B. How to measure national energy sustainability performance: An Icelandic case-study. *Energy Sustain Dev*, 2017, 39, 29–47.

Si, J., Marjanovic-Halburd, L., Nasiri. F., Bell, S. Assessment of building-integrated green technologies: A review and case study on applications of Multi-Criteria Decision Making (MCDM) method. *Sustain Cities Soc*. 2016, 27, 106–15.

Siksnelyte, I., Zavadskas E.K. Achievements of the European Union Countries in Seeking for Sustainable Electricity Sector. *Energies*, 2019, 12(12): 2254.

Siksnelyte, I., Zavadskas, E.K., Bausys, R., Streimikiene, D. Implementation of EU energy policy priorities in the Baltic Sea Region countries: Sustainability assessment based on neutrosophic MULTIMOORA method. *Energy Policy*. 2019, 125, 90–102.

Siksnelyte, I., Zavadskas, E.K., Streimikiene, D., Sharma, D. An Overview of Multi-Criteria Decision-Making Methods in Dealing with Sustainable Energy Development Issues. *Energies*, 2018, 11, 2754.

Siksnelyte-Butkiene, I, Streimikiene, D. Sustainable Development of Road Transport in the EU: Multi-Criteria Analysis of Countries' Achievements. *Energies*, 2022, 15(21), 8291.

Siksnelyte-Butkiene, I., Streimikiene, D., Balezentis, T. Multi-criteria Analysis of Heating Sector Sustainability in Selected North European Countries. *Sustain Cities Soc*, 2021, 69(6), 102826.

Siksnelyte-Butkiene, I., Streimikiene, D., Lekavicius, V., Balezentis, T. Energy poverty indicators: A systematic literature review and comprehensive analysis of integrity. *Sustain Cities Soc*, 2021a, 67, 102756.

Smarandache, F. A. *Unifying Field in Logics: Neutrosophic Logic. Neutrosophy: Neutrosophic Probability, Set and Logistic*, 1999. American Research Press: Rehoboth, DE.

Solangi, Y.A., Tan, Q., Mirjat, N.H., Ali, S. Evaluating the strategies for sustainable energy planning in Pakistan: An integrated SWOT-AHP and Fuzzy-TOPSIS approach. *J Clean Prod*, 2019, 236, 117655.

Sovacool, B.K. Evaluating energy security in the Asia pacific: Towards a more comprehensive approach. *Energy Policy*, 2011, 39(11), 7472–9.

Sovacool, B.K., Mukherjee, I. Conceptualizing and measuring energy security: A synthesized approach. *Energy*, 2011, 36(8), 5343–55.

Spencer, T., Piefederici, R., Waisman, H., Colombier, M., Bertram, C., Kriegler, E., Luderer, G., Humpenöder, F., Popp, A., Edenhofer, O., Elzen, M., Den, Vuuren, D., van, Soest, H., van, Paroussos, L., Fragkos, P., Kainuma, M., Masui, T., Oshiro, K., Akimoto, K., Tehrani, B.S., Sano, F., Oda, J., Clarke, L., Iyer, G., Edmonds, J., Fei, T., Sha, F., Kejun, J., Köberle, A.C., Szklo, A., Lucena, A.F.P., Portugal-Pereira, J., Rochedo, P., Schaeffer, R., Awasthy, A., Shrivastava, M.K., Mathur, R., Rogelj, J., Jewell, J., Riahi, K., Garg, A. *Beyond the Numbers: Understanding the Transformation Induced by INDCs* (No. N°05/15), IDDRI. MILES project, 2015. Paris. www.iddri.org/Publications/Collections/Analyses/MILES%20report.pdf

Spencer, T., R. Pierfederici, O. Sartor, N. Berghmans, S. Samadi, M. Fischedick, K. Knoop, S. Pye, P. Criqui, S. Mathy, P. Capros, P. Fragkos, M. Bukowski, A. Śniegocki, M.R. Virdis, M. Gaeta, K. Pollier, C. Cassisa. Tracking sectoral progress in the deep decarbonisation of energy systems in Europe. *Energy Policy*, 2017, 110, 509–17. http://dx.doi.org/10.1016/j.enpol.2017.08.053

Sperlin, K., Moller, B. End-use energy savings and district heating expansion in a local renewable energy system – a short-term perspective. *Appl Energy*, 2012, 92, 831–42.

Stanujkic, D., Karabasevic, D., Smarandache, F., Zavadskas, E.K., Maksimovic, M. An Innovative Approach to Evaluation of the Quality of Websites in the Tourism Industry: A Novel MCDM Approach Based on Bipolar Neutrosophic Numbers and the Hamming Distance. *Transform Bus Econ*, 2019, 18, 1 (46), 149–62.

Stefaniec, A., Hosseini, K., Assani, S., Hosseini, S.M., Li, Y.J. Social sustainability of regional transportation: An assessment framework with application to EU road transport. *Socio-Econ Plan Sci*, 2021, 78, 101088.

Streimikiene, D., Siksnelyte, I. Sustainability Assessment of Electricity Market Models in selected developed world countries. *Renew Sust Energ Rev*, 2016, 57, 72–82. doi.org/10.1016/j.rser.2015.12.113

Strielkowski, W., Volkova, E., Pushkareva, L., Streimikiene, D. Innovative Policies for Energy Efficiency and the Use of Renewables in Households. *Energies*, 2019, 12, 1392.

Tabatabaei, M.H., Amiri, M., Firouzabadi, S.M.A.K., Ghahremanloo, M., Keshavarz-Ghorabaee, M., Saparauskas. J. A New Group Decision-Making Model based on BWM and its Application to Managerial Problems. *Transform Bus Econ*, 2019, 18, 2 (47), 197–214.

Tattini, J., Gargiulo, M., Karlsson, K. Reaching carbon neutral transport sector in Denmark – Evidence from the incorporation of modal shift into the TIMES energy system modeling framework. *Energy Policy*, 2018, 113, 571–83.

The European Automobile Manufacturers' Association. *ACEA Vehicles in Use Report*. 2021, 21 p. Available at: www.acea.auto/files/report-vehicles-in-use-europe-january-2021-1.pdf (accessed on 15 September 2022).

Thiel, C., Nijs, W., Simoes, S., Schmidt, J., van Zyl, A., Schmid, E. The impact of the EU car CO2 regulation on the energy system and the role of electro-mobility to achieve transport decarbonisation. *Energy Policy*, 2016, 96, 153–66.

Troitino D.R. Transport policy in the European Union. *MEST Journal*, 2015, 3(2), 135–41.

Vavrek, R., Chovancova, J. Assessment of economic and environmental energy performance of EU countries using CV-TOPSIS technique. *Ecol Indic*, 2019, 106, 105519.

Wang, L., Xu, L., Song, H. Environmental performance evaluation of Beijing's energy use planning. *Energy Policy* 2011, 39, 3483–3495.

World Economic Forum, 2015. The Global Energy Architecture Performance Index 2015: Methodological Addendum. www3.weforum.org/docs/WEF_GlobalEnergy ArchitecturePerformance_Index_2015.pdf

Yadoo, A., Cruickshank, H. The role for low carbon electrification technologies in poverty reduction and climate change strategies: A focus on renewable energy mini-grids with case studies in Nepal, Peru and Kenya. *Energy Policy*, 2012, 42, 591–602.

Yang, Y., Ren, J., Solgaard, H.S., Xu, D., Nguyen, T.T. Using multi-criteria analysis to prioritize renewable energy home heating technologies. *Sustain Energy Technol Assess*, 2018, 29, 36–43.

Yılmaz S., Selim H. A review on the methods for biomass to energy conversion systems design. *Renew Sust Energ Rev*, 2013, 25, 420–30.

Zavadskas, E.K., Bausys, R., Juodagalviene, B., Garnyte-Sapranaviciene, I., 2017. Model for Residential House Element and Material Selection by Neutrosophic MULTIMOORA method, *Eng Appl Artif Intell*, 64, 315–24.

Zelazna, A., Golebiowska, J., 2015. The Measures of Sustainable Development – A Study Based on the European Monitoring of Energy-Related Indicators. *Probl Ekorozwoju – Problems of Sustainable Development*, 2015, 10(2), 169–77.

Zhang, C.H., Wang, Q., Zeng, S.Z., Balezentis, T., Streimikiene, D., Alisauskaite-Seskiene, I., Chen, X.L. Probabilistic multi-criteria assessment of renewable micro-generation technologies in households. *J Clean Prod*, 2019a, 212, 582–92.

Zhang, R., Fujimori, S. The role of transport electrification in global climate change mitigation scenarios. *Environ Res Lett*, 2020, 15(3), 034019.

Zhang, T., Shi, X.P., Zhang, D.Y., Xiao, J.J. Socio-economic development and electricity access in developing economies: A long-run model averaging approach. *Energy Policy*, 2019b, 132, 223–31.

4 Multi-Criteria Decision Making for Sustainable Transport Development

4.1 SUSTAINABLE TRANSPORT DEVELOPMENT

Transport sector accounts for more than a quarter of generated energy consumed globally and plays a significant role in terms of sustainable energy development and transition to a low-carbon energy society. In the last decade, major attention has been paid to technological achievements and their application in the market and shared mobility and micro mobility options development. The improvements in technologies allow for more advanced, efficient, often cheaper transport options. But it is necessary to emphasize that sustainable transport development and sustainable transport decision making are not associated with the pollution reduction only. For example, a large share of people, especially those living in metropolises or big cities, suffer from transport noise and pollution. Road traffic is the main noise source, which is followed by noise caused by trains, planes and industry. Road transport is distinguished not only as the main contributor of transport noise, but also as a major source of greenhouse gas (GHG) emissions. It can be stated that the main air pollution source in cities is public and private transport. It follows that the transport sector, especially road transport, has a huge impact on human health (Tang et al., 2020; Spadaro and Pirlone, 2021) and quality of life, and the state of the environment. In the context of climate change, and with the recognition of the impact of transport on human health and the environment, decision makers must develop the transport sector in more sustainable ways and use solutions more friendly to the environment. However, simultaneously it is important to ensure affordability, reliability, and convenience of transport systems. Many scientists argue that it is important to follow the concept of sustainability in developing future transport systems. For example, Cornet et al. (2018) stressed the significance of sustainability in transport studies and proposed a methodology for the sustainable point of view. A methodological framework, which helps to identify and select indicators for sustainable transport development, was presented by Castillo and Pitfield (2010). As stressed by Cadena et al. (2015), it is very important that criteria selected would reflect not only the opinion of the experts, but also would reflect social and geographical contexts of the project implemented. This can be done by including various groups of stakeholders in the decision making processes.

Decision making in the transport sector is complicated not only due to different characteristics of alternatives and possible solutions, but also due to multidimensional requirements of transport service providers, passengers, freight companies, and decision makers themselves. Transport development issues always are linked and impact many different groups of stakeholders. Therefore, it is very important that various stakeholders would be involved in decision making regarding issues related to transport development. Only then can the optimal solution be found, and the needs of all groups of stakeholders met as much as possible.

Recent studies emphasize the necessity of stakeholders' participation in sustainable transport decision making (for example, Barfod, 2018; Huang et al., 2021 Rzesny-Cieplinska et al., 2021; Lee et al., 2021; Karolemeas et al., 2021). The engagement of various stakeholders brings multiple benefits, such as: democratizing decision making, raising awareness, sharing knowledge and information, encouraging discussions, and so forth. But perhaps the most important thing is that it allows for looking at the problem from different perspectives. Stakeholder engagement can serve as a main instrument in the process of indicators selection and/or determination of their weights. It also can serve as instrumentally affecting the responsibility to accept and implement the final outcome of the project. As there are various groups of stakeholders in the transport sector, and which have different priorities, preferring different alternatives, multi-criteria analysis is a suitable instrument to consider and evaluate possible contradictory solutions and to select the most preferable one.

4.2 SELECTION OF INSTRUMENTS FOR SUSTAINABLE TRANSPORT DECISION MAKING

Sustainable decision making regarding transportation issues began to be very popular in the last decade. The analysis of scientific studies in the Web of Science database (2022) showed that there is a significant increase in studies dealing with both transport sustainability and multi-criteria analysis in the last decade. Also, a clear trend of increase each year is notable. More than 16,500 publications can be found on the topic "sustainable transport" in the database, published in the period 1991–2021. It can be stated that multi-criteria analysis is a very popular tool in different decision making areas, and one of the main instruments used to deal with energy sustainability questions. More than 22,000 studies have been published on the topics "multi-criteria decision making" or "multi-criteria decision analysis" in the Web of Science database (Figure 4.1). Also, the popularity of studies using multi-criteria decision making (MCDM) instruments for the sustainable decision making in transport have significantly increased recently (Figure 4.2).

Figure 4.3 shows keywords and their frequency of use in studies selected in the Web of Science database of different combination of the topics "sustainable transport" and "multi-criteria decision making" or "multi-criteria decision analysis". The variety of questions and problems in sustainable transport decision making is also seen in Figure 4.3. This keywords map also displays supporting methods and instruments used for the studies performed. Also, it reveals the problem of terms inconsistency in the scientific literature.

Different MCDM techniques have been applied to deal with various transport sustainability questions. But the most popular can be easily singled out. The most widely applied MCDM method is the Analytical Hierarchy Process (AHP), developed by Saaty (Saaty, 1980). Also, among the quite popular are such techniques as Preference

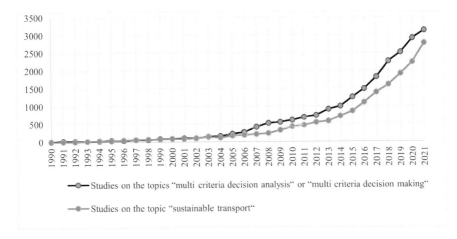

FIGURE 4.1 Number of studies on the different topics.

Source: Data from the Web of Science database (2022).

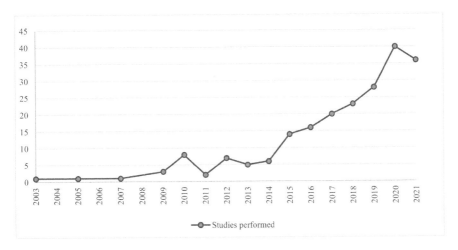

FIGURE 4.2 Number of studies dealing with sustainable transport development by the application of MCDM techniques.

Source: Data from the Web of Science database (2022).

Ranking Organization Method for Enriching Evaluation (PROMETHEE) introduced by Mareschal and Brans (1992); the Technique for Order of Preference by Similarity to the Ideal Solution (TOPSIS), presented by Hwang and Yoon (1981); and Simple Additive Weighting (SAW), introduced by MacCrimon (1968). Other applied instruments can be mentioned, including such methods as, the Simple Multi Attribute Rating Technique (SMART) (Edwards, 1977) and the Simple Multi Attribute Rating Technique Exploiting Ranks (SMARTER) (Edwards and Barron, 1994); the Best

FIGURE 4.3 Variety and popularity of keywords in studies dealing with sustainable transport decision making.

Source: Based on data from the Web of Science database (2022).

Worst Method (BWM) (Rezaei, 2015); and the Multicriteria Optimization and Compromise Solution (VIKOR) (Opricovic, 1998). Some rarely used methods can be singled out: Analytic Network Process (ANP) (Saaty, 1996); Decision-Making Trial and Evaluation Laboratory (DEMATEL) (Gabus and Fontela, 1972); Elimination and Choice Transcribing Reality (ELECTRE) (Roy, 1968; Vallee and Zielniewicz, 1994); Multi-Attribute Border Approximation Area Comparison (MABAC) (Pamucar and Cirovic, 2015); Additive Ratio Assessment (ARAS) (Zavadskas and Turskis, 2010); the Characteristic Objects Metod (COMET) (Salabun, 2014); The Evaluation Based on Distance from Average Solution (EDAS) (Keshavarz-Ghorabaee et al., 2015); Multi-Attribute Utility Theory (MAUT) (Keeney, 1982; Raiffa, 1970); Multiple-Attribute Value Theory (MAVT) (Fishburn, 1967; Keeney and Raiffa, 1976).

Among the very rarely applied methods to deal with transport sustainability issues are the following: Combined Compromise Solution (CoCoSo) (Yazdani et al., 2019); Combinative Distance-based Assessment (CODAS) (Keshavarz-Ghorabaee et al., 2016); COPRAS (Complex Proportional Assessment) (Zavadskas et al., 1994); Full Consistency Method (FUCOM) (Pamucar et al., 2018); Measuring Attractiveness by a Category Based Evaluation Technique (MACBETH) (Bana e Costa et al., 2012); MAJA (Jacyna, 2006); Multiplicative Exponential Weighting (MEW) (Zanakis et al. 1998); the Integrated Value Model for Sustainability Assessment (MIVES) (San-Jose and Cuadrado, 2010); Full Multiplicative Form of Multi-Objective Optimization by

Ratio analysis (MULTIMOORA) (Brauers and Zavadskas, 2010); Novel Approach to Imprecise Assessment and Decision Environments (NAIADE) (Munda, 1995); Sequential Interactive Modelling for Urban Systems (SIMUS) (Munier, 2016); Step-Wise Weight Assessment Ratio Analysis (SWARA) (Kersuliene et al., 2010); Interactive Multi-criteria Decision Making (TODIM) (Gomes and Lima, 1992); Weighted Sum Method (WSM) (Zadeh, 1963).

It necessary to mention that many supporting methods have been combined with MCDM approaches to deal with sustainable transport decision making. As the most popular can be singled out, such methods as: Fuzzy Set Theory; Cost-Benefit Analysis; Life-Cycle Assessment; DEA; System Dynamics Simulation; Geographic Information System (GIS); Multi Actor Multi-Criteria Analysis (MAMCA); Quality Function Deployment; Prospect Theory; Monte Carlo Simulation; Goal Programming; different methods for determination of criteria weights (e.g., Pairwise Comparison, Delphi Method, Swing Weights, Kendall's Concordance Coefficient).

Interest in sustainable transport decision making started to increase in the last decade in both political documents and scientific literature. The application of different MCDM techniques for sustainable transport decision making started to be one of the most popular instruments, which allows for accounting contradictory aspects and different needs of various stakeholders. Analysis of scientific literature allowed for singling out the main application areas, where different MCDM methods has been applied, and to distinguish the most popular techniques in each category. The main transport studies categories and the most popular methods used for sustainable decision making are presented in Figure 4.4.

The largest number of studies have been performed in the public transport planning category. Various different methods have been used in this category. Studies dealing with public transport planning issues are usually designed to help policy makers and local authorities and serve as supporting tools for decision making. The studies carried out address various sustainable transport planning issues, such as: selection of transport, identification of park-and-ride locations, planning new transport infrastructure, management of existing transport infrastructure, and so forth.

Studies aimed to measure sustainability are performed for many different purposes, such as: transport infrastructure assessment, transport services assessment, evaluation of transport sector as a whole or for a specific city or region, and so forth.

As for public transport planning studies, studies dealing with transport policy issues are also designed as supporting tools for policy makers and local authorities. Studies in this category are oriented towards implementation of transport and energy policy objectives.

The decision making problems analyzed in the transport project selection category vary from the assessment systems for the analysis and ranking investment projects to evaluation of possible mobility alternatives. Issues in the logistics category are usually dedicated to sustainable freight planning.

The variety of methods applied in each category is presented in Table 4.1. It should be noted that this list is not exhaustive. The application of MCDM methods to solve transport sustainability issues is becoming more and more popular and different additional techniques will be applied in the future. Also, some of the studies can

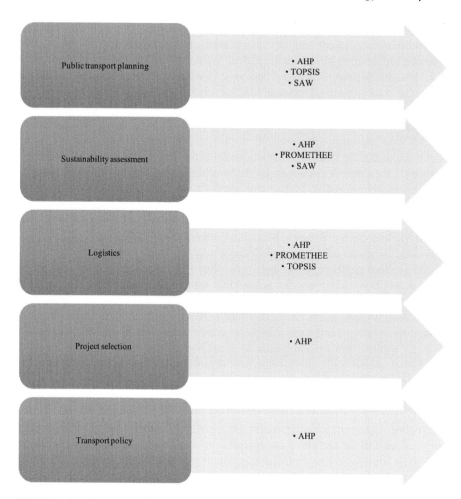

FIGURE 4.4   The main MCDM application areas and the most popular methods used for the sustainable transport decision making.

be classified into several categories. Therefore, it is necessary to mention that the following classification is subjective.

MCDM methods to solve decision making problems for transport planning and sustainability evaluation are quite popular tools. The popularity of methods varies among categories, but one technique can be singled out as the most popular in all categories (Figure 4.4): the AHP technique is the most widely applied, not only in sustainable transport decision making, but also as the most widely applied technique to solve various energy sustainability problems in many areas. TOPSIS can be mentioned as a popular approach in public transport planning and logistics categories. The PROMETHEE method has been applied many times in sustainability assessment and logistics categories. SAW technique is quite popular for sustainability assessment studies.

TABLE 4.1
Variety of Methods Applied and Their Popularity in Sustainable Transport Decision Making

	Public transport planning	Sustainability assessment	Logistics	Project selection	Transport policy
AHP	Often	Often	Often	Often	Often
PROMETHEE	Rarely	Often	Often	Very rarely	Rarely
TOPSIS	Quite often	Very rarely	Quite often	Rarely	
SAW	Quite often	Often	Very rarely		
SMART/SMARTER	Rarely	Very rarely		Rarely	
BWM	Rarely	Very rarely		Very rarely	
VIKOR	Very rarely	Very rarely	Very rarely		
ANP				Rarely	Very rarely
DEMATEL	Very rarely		Very rarely	Very rarely	
ELECTRE III		Quite often			
MABAC	Rarely		Very rarely		
ARAS	Rarely				
COMET	Very rarely				
EDAS	Very rarely		Very rarely		
MOUT	Very rarely				Very rarely
MAVT		Very rarely			Very rarely
Others	Very rarely	Very rarely	Very rarely	Very rarely	Very rarely

Source: Created by authors.

4.2.1 PUBLIC TRANSPORT PLANNING ISSUES

Many different issues have been addressed in studies dealing with sustainable transport planning. These studies can be listed under several main categories.

One part of the studies addresses decision making problems related to mobility choices. The multi-criteria BWM technique was applied for the identification changes of mobility choices during the COVID-19 pandemic in the study by Moslem et al. (2020). The presented case study in two Italian cities (Catania and Palermo) revealed that during the pandemic a tendency to walk short distances appeared, and the use of public transport services significantly decreased. Bulckaen et al. (2016) sought to evaluate mobility choices taking into account preferences of stakeholders and different sustainability issues. The authors proposed a framework oriented towards a synergy among needs of stakeholders and sustainability issues and applied PROMETHEE and MAMCA approaches. The proposed framework was applied in practice for the three case studies. According to the results, the preferences of stakeholders were oriented towards sustainable options. AHP technique was used by Bivina and Parida (2020) for the identification of pedestrians' priorities in a city in India. The results showed

that the most important thing for city pedestrians is safety of the walking environment. BMW and AHP techniques were used to plan park-and-ride locations in the
study by Ortega et al. (2020). The authors identified that "accessibility of public
transportation" is the most important factor when selecting park-and-ride locations.

Other research focuses on questions related to selection of public transport
alternatives (solutions for roadway intersections, bicycle development, railway traffic
problems, management solutions, etc.). A framework for the sustainability measurement of roadway intersections was presented in the study by Al-Kaabi et al. (2020).
The framework focuses on road-users and follows the traditional concept of sustainability. The authors applied TOPSIS technique for the sustainability assessment of
four intersections in the United Arab Emirates. Shishegaran et al. (2020) sought to
determine the most preferable scenario for the improvement of traffic conditions in a
Tehran (Iran) interchange. The authors evaluated six possible scenarios and used the
TOPSIS method for the ranking.

Several studies focus on sustainable bicycle environment planning. Salabun et al.
(2019) addressed the issues of sustainable city planning, using COMET technique
for identification of the most preferable bicycle option. The presented methodology
allows for considering alternatives under incomplete knowledge conditions. A hybrid
model based on SAW and GIS approaches was introduced by Rybarczyk and Wu
(2010) for the bicycle facility planning in a city of the United States (Milwaukee).
A tool for the ranking of renewal priorities of bicycle pathways was introduced by
Zagorskas and Turskis (2020a). The proposed methodology relies on combination of
ARAS and GIS approaches.

The framework for the measurement of railway traffic safety was proposed by
Blagojevic et al. (2020). Eleven criteria were selected that reflect different railway
traffic sustainability issues. The authors presented a case study, where MABAC technique was applied for the measurement of safety level in Bosnia and Herzegovina
railways. Another study for sustainable railways development was presented by
Stoilova et al. (2020). The authors introduced a methodology for the assessment and
classification of network performance of railways. The created methodology has been
applied for the Trans-European Transport Network corridor assessment. The SIMUS
method was used for calculations and countries ranking.

Other studies are oriented towards management solutions. For example, Yang
et al. (2016) combined ANP, DEMATEL, and Goal Programing to create a tool for
more sustainable transport infrastructure planning. The proposed tool includes environmental costs into consideration. MAJA approach was used in the study Ciesla
et al. (2020) for the development of a model for the sustainable transport planning
in metropolises. Ecological, safety, economical and qualitative aspects of transport
systems were included in the model to ensure sustainable transport system planning.
Pamucar et al. (2020) presented a model for transportation demand management.
The proposed model relies on the FUCOM-D'Bonferroni approach and is designed
as a supporting tool for transport planners and local authorities to deal with mobility
system management. The validity of the model is proved in comparison with others
MCDM methods. The model was applied for a case study in Istanbul (Turkey). GIS,
SAW, TOPSIS methods were applied in the study by Jakimavicius and Burinskiene

(2009) to develop a framework for transportation zones analysis and ranking in Vilnius (Lithuania). To deal with a growing number of cars in a city, Palevicius et al. (2016) presented a framework that combines TOPSIS, PROMETHEE, and SAW techniques. The authors examined 49 parking lots of different shopping centers in Vilnius and evaluated how to help solve parking problems in the city. A prioritization model for public transport systems failure analysis was introduced by Erdogan and Kaya (2019). The calculations are based on AHP and TOPSIS, where AHP is used for weighting and TOPSIS is applied to rank the alternatives. Ma et al. (2019) sought to evaluate quality of bike sharing services. The case study in one city of China was performed combining DEMATED and VIKOR techniques. The main attention was paid to different groups of stakeholders in the assessment: users, local authorities, operators of sharing platforms, and bike associations. Casanovas-Rubio et al. (2020) developed a tool for the assessment of the impact of construction works on mobility. The proposed tool is dedicated for the local authorities and construction planners.

It is worth emphasizing that many studies pay significant attention to the role of stakeholders in the process of sustainable transport decision making. Some studies are dedicated to create methodologies for successful stakeholders' engagement for sustainable public transport planning. Such methodologies allow for awareness of different needs and preferences among different groups of stakeholders and provide guidelines how to find consensus. The most proposed technique in created methodologies is AHP. Examples of application of the AHP approach are methodologies created by Duleba and Moslem (2018), Moslem et al. (2019), Ghorbanzadeh et al. (2019).

4.2.2 SUSTAINABILITY ASSESSMENT

The decision making questions addressed in the sustainability assessment category varies from infrastructure development or analysis of transport services to the measurement of sustainability of the transport sector as a whole.

It can be stated that in this category the largest number of studies are dedicated to dealing with sustainable infrastructure development. The combinations of different multi-criteria methods have been applied to evaluate the sustainability of infrastructure reducing transportation noise. Oltean-Dumbrava et al. (2013) developed a tool that helps to make sustainable decisions regarding the reduction of transport noise. The created methodology is based on a combination of AHP, SAW, SMART, and WSM methods. The sustainability assessment of roadside noise barriers was performed in the study by Oltean-Dumbrava and Miah (2016). The authors used a combination of ELECTRE III, PROMETHEE, and SAW techniques and proposed the methodology that can be used for the assessment of existing and new roadside noise barriers assessment. Also, the proposed methodology was applied in the next studies (Oltean-Dumbrava et al., 2016): Zapolskyte et al. (2020) combined AHP, COPRAS, SAW, and TOPSIS methods in order to measure sustainability of transport infrastructure and transport services in Vilnius. Sustainability of transport services was assessed also by Alzouby et al. (2019). The authors focused on accessibility issues of disabled people and performed analysis in a city in Jordan by the application of the AHP technique for calculations. Transportation service providers and its sustainability were discussed in the study by Paul et al. (2020). The authors proposed a framework that reflects economic, environmental, social, and operational factors and combines BMW and VIKOR techniques for calculations. A framework for the urban transport projects sustainability assessment based on the AHP method was presented by Jones et al. (2013). The framework allows for evaluating transport projects and reflects local issues. The authors presented a case study of Accra (Ghana).

Other studies deal with transport sector sustainability assessment as a whole. Some studies are developed for city-level assessment, while others are developed for country-level. For example, Jasti and Ram (2019a) selected 29 indicators and applied AHP technique for the sustainability evaluation of public transport system and metro rail system in selected cities of India (Jasti and Ram, 2019b). A sustainability index for transport system assessment in a city of Spain (San Sebastian) was presented by Oses et al. (2017). The authors applied the MIVES method for computation processes. A large-scale study was performed by Shmelev and Shmeleva (2018), where 57 cities of different countries were assessed in terms of dimensions of a smart city. The authors created a set from 20 indicators, which include also those that specify conditions of public transportation, mobility, and cycling in a city. APIS, ELECTRE III, and NAIADE methods were used for calculations. The analysis cities in Poland and their compliance with the criteria of smart city were in the study by Ogrodnik (2020). The author selected 43 indicators, reflecting the concept of smart city and applied PROMETHEE for calculations.

Country-level transport sector sustainability assessment was performed in a study by Wang et al. (2022). The authors carried out analysis and measured sustainability of road transportation systems in the countries of the Organization for Economic Cooperation and Development (OECD). The integrated entropy-CoCoSo technique was applied for calculations and country rankings. Siksnelyte-Butkiene and Streimikiene (2022) created an indicators system for road transport sustainability assessment in the EU member states. The TOPSIS technique was applied for calculations and country rankings.

Other studies focus on impact assessment. For example, the impact of end-of-life tires on expenditures of transportation and the level of emissions of vehicles was measured in the study by Nowakowski and Krol (2020). The combination of two multi-criteria techniques, AHP and PROMETHEE, was applied for calculations. Sustainability assessment of fossil- and biomass-based transportation fuels was performed in the study by Ekener et al. (2018). The authors developed a methodology based on life cycle assessment and MAVT methods.

4.2.3 Logistics Issues

It can be stated that decision making problems in the logistics category are oriented towards sustainable freight planning. For this purpose, many different techniques have been applied. Also, exclusively in this category, the developed assessment and decision making systems pay much attention to the engagement of stakeholders in the decision making processes in order to reflect the needs of different groups of stakeholders. For example, AHP and CODAS methods were used for determination of the most preferable place for dry port terminals (Tadic et al., 2020). The developed model pays attention to stakeholders' requirements and reflects aspects related to economic, ecological, and social dimensions of the problem analyzed. The developed model was applied for a Balkans region case study. The expectations of customers regarding port services were analyzed in a study by Wang et al. (2017). The authors created a model for sustainable decision making for transport services, one which is based on AHP technique. AHP and PROMETHEE methods were applied in the study by Aljohani and Thompson (2019) in order to choose the most preferable delivery fleet alternative. The authors emphasize in their study how important is to engage stakeholders in decision making when dealing with issues related to urban freight policy. Issues related to sustainable logistic in urban areas were analyzed also in the study by Semanjski and Gautama (2019). The authors provided a model for stakeholders' engagement, which is based on the AHP method. The electric freight vehicles for logistics in urban areas were assessed by Watrobski et al. (2017). TOPSIS and PROMETHEE were applied for the calculations and vehicles ranking.

Although a lot of different methods have been applied for sustainable freight planning, the most popular is AHP and PROMETHEE. For the sustainable selection of freight alternatives, Simongati (2010) applied SAW and PROMETHEE. Vermote et al. (2013) combined MAMCA and AHP approaches to develop a decision making system to plan freight route networks in the Flanders region of Belgium. Four freight rings were considered, taking into account preferences of various stakeholders. The results showed that freight companies are oriented towards local accessibility, while the local government and citizens require safety of traffic and good living conditions. AHP and DEMATED were used for calculations in the study by Kijewska et al. (2018) in order to solve delivery questions and rationalize freight flow in a region of Poland. In order to ensure distribution services in a sustainable way, Macharis and Milan (2015) presented a methodology which relies on stakeholders' participation in decision making processes, and a MAMCA approach and multi-criteria AHP and PROMETHEE methods for ranking the alternatives. The extended TOPSIS approach was applied in the study by Chen et al. (2019), where a framework to select hazardous materials transportation solutions was created. The problem of sustainable global supplier selection was addressed in the study by Awasthi et al. (2018). The authors developed a framework and selected criteria for the evaluation which reflect economic, environmental, social, quality, and global risk aspects. AHP and VIKOR methods were used for calculations.

The other part of the studies dealing with sustainable logistics issues assess sustainability of transportation systems. For example, Yazdani et al. (2020) sought to evaluate the performance of freight transport systems. The authors presented a framework that follows the traditional concept of sustainability, and applied DEMATEL and MABAC techniques for seven freight companies' assessment. The results were also calculated and compared with CODAS, EDAS, and TOPSIS methods. According to the authors, the methods based on rough numbers have advantages compared with fuzzy or interval based ones. Environmental responsibility of providers of freight transport services was evaluated by Kumar and Anbanandam (2020). The authors created a methodology that relies on AHP and VIKOR methods and applied it for the freight transport sector assessment in India. In order to evaluate sustainability of transport fleets, Bai et al. (2017) created a framework considering environmental, economic, and vehicle performance indicators. The calculations and ranking of the alternatives were performed based on the VIKOR method.

4.2.4 Transport Project Selection Issues

The decision making questions addressed in transport project selection are related to many issues, but in general the two main groups can be singled out, which are selection of the most preferable transport development alternative and consideration of solutions regarding the investments.

Studies dealing with sustainable project selection regarding transport infrastructure development are prevalent in this category. Generally, the evaluation frameworks proposed and the case studies carried out are dedicated to large-scale projects and can serve as a supporting tool in the decision making process. For example, Barfod and Sailing (2015) presented a framework that considers sustainability and economic and strategic perspectives. The proposed framework is based on AHP, SMART, SMARTER techniques and was applied to find the best option for a new, fixed link between Denmark and Sweden. AHP, PROMETHEE, and TOPSIS were used for calculations in the study by Broniewicz and Ogrodnik (2020) in order to find the most preferable option for the expressway section in Poland. Pryn et al. (2015) introduced a decision support system called SUSTAIN, which is oriented towards sustainable decision making and is dedicated to decision makers dealing with transport projects. The methodology is based on AHP and SMARTER methods and was applied for two case studies in Frederikssund (Denmark), planning connection crossing construction and solving bridge congestion issues (Salling and Pryn, 2015). Several multi-criteria techniques were combined in the study by Zagorskas and Turskis (2020b). The authors sought to create a new model, which can help to identify the most suitable locations for pedestrian bridges. The presented hybrid model combines multi-criteria ARAS, EDAS, MEW, SWARA MCDM techniques and the GIS approach. Turskis et al. (2019) applied MULTIMOORA in order to find the most suitable option for a second runway for the airport in Vilnius. Creating a project selection system Tadic et al. (2019) took into account the needs of different stakeholders and followed the concept of sustainability. The authors applied ANP for the calculations in order to find the most suitable solution for planning intermodal terminals. Macharis et al. (2012) combined the MAMCA with AHP and PROMETHEE techniques in order to develop a tool for transport project assessment. The presented methodology focuses on the preferences and requirements of stakeholders. Donais et al. (2019) emphasized the importance of collaboration with professionals in the decision making process. The authors created a framework that is sustainability oriented and dedicated to finding and selecting streets for redesign in a city of Canada. The MACBETH technique was used for calculations.

Several evaluation systems can be found in the literature that have been developed in order to find the best solution for investments in the transport sector. For example,

Henke et al. (2020) created an evaluation system that combines the AHP method and cost–benefit analysis; Mohagheghi et al. (2017) introduced a methodology based on the TODIM method. Van de Kaa et al. (2017) sought to evaluate which vehicle alternative has the greater potential to be widely developed in the market. The authors analyzed two alternatives: hydrogen fuel cell powered vehicles and battery powered electric vehicles. The results showed that technological superiority, reputation, and brand credibility are the main factors influencing the success of vehicles in the market, and battery powered electric vehicles have greater potential to be widely developed.

4.2.5 TRANSPORT POLICY ISSUES

Studies dealing with transport policy issues are mostly dedicated to policy makers or local authorities as supporting instruments for decision making processes. The questions analyzed in this category focus on achievement of policy objectives. Mainly, the studies focus on the implementation of climate change goals. While the others pay attention to the participation and transparency of citizens in policy-making processes.

For example, various energy policy scenarios to implement low-carbon objectives in Bangkok were modelled and assessed in the study by Phdungsilp (2010). The results revealed that the transport sector has the biggest influence for the development of a low carbon city, and the shift from private passenger cars to public transport has the biggest potential to reduce emissions. Different energy efficiency measures for transport and building sectors in Greece were evaluated in the study by Neofytou et al. (2020). The measures for the reduction of GHG emissions were assessed by the application of the PROMETHEE II method. According to the results, the measures regarding the improvement in energy efficiency in the building sector are more effective than in transport. A holistic approach for strategic transport planning in urban areas was presented by Kramar et al. (2019). The approach focuses on achievement of sustainable policy objectives and is based on AHP technique. Sayyadi and Awasthi (2020) assessed five transportation policies scenarios to find the most sustainable option. The authors used a dynamics simulation model system and ANP technique. The results showed that trip sharing policy is the most sustainable alternative. Hofer and Madlener (2020) evaluated four energy transition scenarios and applied MOUT technique for ranking. Although the developed model was applied, not only in transport, but also in the other sectors of energy, the methodology proposed can be adapted for decision making in one sector as well. Ullah et al. (2018) presented a decision making framework for road policy planning. The proposed framework developed to consider different gaseous alternatives for road transport in Pakistan relies on the AHP method. A methodological framework for sustainable transport development in the future was presented by Soria-Lara and Banister (2018). The authors performed a case study in a region of Spain, and the AHP method was used for calculations. According to Corral and Hernandez (2017), attention should mainly be paid to the engagement of various stakeholders in the transport planning process, because it is the main condition of sustainable planning. The authors presented a case study for the Canary Islands and used the NAIADE method for the sustainable transport policy planning.

4.3   CRITERIA FOR SUSTAINABLE TRANSPORT DECISION MAKING

Sustainable transport development depends on how the concept of sustainability is followed in the decision making process. Therefore, it is very important to ensure that the criteria selected for decision making reflect all dimensions of sustainability, consider the specific location characteristics, reflect geographical and social context related to the questions analyzed, and consider the preferences of various groups of stakeholders. Stakeholder engagement is desirable in all processes of decision making, but most important is to involve them in the criteria selection and weighting stage. Also, stakeholder engagement can serve as a useful instrument to share knowledge and information and to raise public awareness and acceptability regarding transport development issues. It is also very important to select appropriate methods for evaluating and ranking the alternatives. Each method has its advantages and disadvantages. Also, before selecting the method, it is important to pay attention to what kind of criteria can be used for calculations, that is, whether there can be negative values, values in intervals, large deviations in values, qualitative data, and so forth.

The development of a set of representative criteria for sustainable decision making in the transport sector is necessary not only for future scientific papers, but also for practitioners. Although the instruments to support decision making are essential for policy makers who seek to implement goals of sustainable development, often the essence of sustainability is ignored. In many cases, not enough attention is paid to the environmental and social aspects in decision making. To fill this gap, the thematic areas for criteria selection are provided (Table 4.2). It allows for considering the most important issues in sustainable transport decision making.

Although, the multi-criteria analysis allows for addressing various conflicting issues at various levels, the process of criteria and indicators selection for the assessment is subjective. In order to deal with subjectivity, the usage of additional

TABLE 4.2
Thematic Areas for Criteria Selection

Economic	Environmental	Social
• Project cost / cost of the alternative	• Energy consumption	• Safety criteria
• Maintenance cost	• Usage of resources	• People health
• Operating cost	• Share of RES	• Quality of life (such criteria as: noise level, visual impact, disruption, etc.)
• Time cost	• GHG emissions	• Impact on employment
• Risk / Accident cost	• Air quality / air pollution	• Impact on businesses and community services
	• Noise level	
	• Impact on species	
	• Landscape degradation	
	• Other impact on the environment	

Source: Created by authors.

methods for criteria selection and determination of their weights is desirable. For example, such methods as focus groups, Delphi procedure, and semi-structured interviews are suitable instruments in order to identify opportunities or challenges for sustainable transport development. It also allows for considering stakeholders' priorities, requirements, and acceptability.

REFERENCES

Aljohani, K., Thompson, R.G. A Stakeholder-Based Evaluation of the Most Suitable and Sustainable Delivery Fleet for Freight Consolidation Policies in the Inner-City Area. *Sustainability*, 2019, 11(1), 124.

Al-Kaabi, M.J., Maraqa, M.A., Hawas, Y.S. Development of a Composite Sustainability Index for Roadway Intersection Design Alternatives in the UAE. *Sustainability*, 2020, 12(20), 8696.

Alzouby, A.M., Nusair, A,A., Taha, L.M. GIS based Multi Criteria Decision Analysis for analyzing accessibility of the disabled in the Greater Irbid Municipality Area, Irbid, Jordan. *Alex Eng J*, 2019, 58(2), 689–98.

Awasthi, A., Govindan, K., Gold, S. Multi-tier sustainable global supplier selection using a fuzzy AHP-VIKOR based approach. *Int J Prod Econ*, 2018, 195, 106–17.

Bai, C.G., Fahimnia, B., Sarkis, J. Sustainable transport fleet appraisal using a hybrid multi-objective decision making approach. *Ann Oper Res*, 2017, 250(2), 309–40.

Bana e Costa, C.A., De Corte, J.M., Vansnick, J.C. MACBETH. *Int J Inf Technol Decis Mak*, 2012, 11(2), 359–87.

Barfod, M.B. Supporting Sustainable Transport Appraisals Using Stakeholder Involvement and MCDA. *Transport*, 2018, 33(4), 1052–66.

Barfod, M.B., Sailing, K.B. A new composite decision support framework for strategic and sustainable transport appraisals. *Transp Res A: Policy Pract*, 2015, 72, 1–15.

Bivina, G.R., Parida, M. Prioritizing pedestrian needs using a multi-criteria decision approach for a sustainable built environment in the Indian context. *Environ Dev Sustain*, 2020, 22(5), 4929–50.

Blagojevic, A., Stevic, Z., Marinkovic, D., Kasalica, S., Rajilic, S. A Novel Entropy-Fuzzy PIPRECIA-DEA Model for Safety Evaluation of Railway Traffic. *Symmetry-Basel*, 2020, 12(9), 1479.

Brauers, W.K.M., Zavadskas E.K. Project Management by MULTIMOORA as an Instrument for Transition Economies. *Technol Econ Dev Econ*, 2010, 16(1), 5–24.

Broniewicz, E., Ogrodnik, K. Multi-criteria analysis of transport infrastructure projects. *Transp Res D Transp Environ*, 2020, 83, 102351.

Bulckaen, J., Keseru, I., Macharis, C. Sustainability versus stakeholder preferences: Searching for synergies in urban and regional mobility measures. *Res Transp Econ*, 2016, 55, 40–49.

Cadena, P.C.B., Magro, J.M.V. Setting the Weights of Sustainability Criteria for the Appraisal of Transport Projects. *Transport*, 2015, 30(3), 298–306.

Casanovas-Rubio, M.D., Ramos, G., Armengou, J. Minimizing the Social Impact of Construction Work on Mobility: A Decision-Making Method. *Sustainability*, 2020, 12(3), 1183.

Castillo, H., Pitfield, D.E. ELASTIC – A methodological framework for identifying and selecting sustainable transport indicators. *Transp Res D Transp Environ*, 2010, 15(4), 179–88.

Chen, Z.S., Li, M., Kong, W.T., Chin, K.S. Evaluation and Selection of HazMat Transportation Alternatives: A PHFLTS- and TOPSIS-Integrated Multi-Perspective Approach. *Int J Environ Res Public Health*, 2019, 16(21), 4116.

Ciesla, M., Sobota, A., Jacyna, M. Multi-Criteria Decision Making Process in Metropolitan Transport Means Selection Based on the Sharing Mobility Idea. *Sustainability*, 2020, 12(17), 7231.

Cornet, Y., Barradale, MJ., Barfod, MB., Hickman, R. Giving current and future generations a real voice: A practical method for constructing sustainability viewpoints in transport appraisal. *Eur J Transp Infrastruct Res*, 2018, 18(3), 316–39.

Corral, S., Hernandez, Y. Social Sensitivity Analyses Applied to Environmental Assessment Processes. *Ecol Econ*, 2017, 141, 1–10.

Donais, F.M., Abi-Zeid, I., Waygood, E.O.D., Lavoie, R. Assessing and ranking the potential of a street to be redesigned as a Complete Street: A multi-criteria decision aiding approach. *Transp Res Part A Policy Pract*, 2019, 124, 1–19.

Duleba, S., Moslem, S. Sustainable Urban Transport Development with Stakeholder Participation, an AHP-Kendall Model: A Case Study for Mersin. *Sustainability*, 2018, 10(10), 3647.

Edwards, W. How to use multi-attribute utility measurement for social decision-making. *IEEE Trans Syst Man Cybern*, 1977, 7, 326–40.

Edwards, W., Barron, F.H. SMARTS and SMARTER: Improved Simple Methods for Multiattribute Utility Measurement. *Organ Behav Hum Decis Process*, 1994, 60, 306–25.

Ekener, E., Hansson, J., Larsson, A., Peck, P. Developing Life Cycle Sustainability Assessment methodology by applying values-based sustainability weighting – Tested on biomass based and fossil transportation fuels. *J Clean Prod*, 2018, 181, 337–51.

Erdogan, M., Kaya, I. Prioritizing failures by using hybrid multi criteria decision making methodology with a real case application. *Sustain Cities Soc*, 2019, 45, 117–30.

Fishburn, P.C. Methods of estimating additive utilities. *Manage Sci*, 1967, 13, 435–53.

Gabus, A., Fontela, E. World Problems, An Invitation to Further Thought within the Framework of DEMATEL. Battelle Geneva Research Centre, Geneva, 1972, 1–8.

Ghorbanzadeh, O., Moslem, S., Blaschke, T., Duleba, S. Sustainable Urban Transport Planning Considering Different Stakeholder Groups by an Interval-AHP Decision Support Model. *Sustainability*, 2019, 11(1), 9.

Gomes, L., Lima, M. TODIM: Basics and application to multicriteria ranking of projects with environmental impacts. *Found Comput Decis Sci*, 1992, 16, 113–27.

Henke, I., Carteni, A., Di Francesco, L. A Sustainable Evaluation Processes for Investments in the Transport Sector: A Combined Multi-Criteria and Cost-Benefit Analysis for a New Highway in Italy. *Sustainability*, 2020, 12(23), 9854.

Hofer, T., Madlener, R. A participatory stakeholder process for evaluating sustainable energy transition scenarios. *Energy Policy*, 2020, 139, 111277.

Huang, H., De Smet, Y., Macharis, C., Doan, NAV. Collaborative decision-making in sustainable mobility: Identifying possible consensuses in the multi-actor multi-criteria analysis based on inverse mixed-integer linear optimization. *Int J Sustain Dev World Ecol*, 2021, 28(1), 64–74.

Hwang, C.L., Yoon, K. *Multiple Attributes, Decision Making Methods and Applications*. Springer: Berlin, Heidelberg, 1981, 22–51.

Jacyna, M. The Multiobjective Optimisation to Evaluation of the Infrastructure Adjustment to Transport Needs. In K.G. Goulias (Ed.), *Transport Science and Technology*. Emerald Group Publishing, Bingley, 2006, 395–405.

Jakimavicius, M., Burinskiene, M. A GIS and Multi-Criteria-Based Analysis and Ranking of Transportation Zones of Vilnius City. *Technol Econ Dev Econ*, 2009, 15(1), 39–48.

Jasti, P.C., Ram, V.V. Integrated and Sustainable Benchmarking of Metro Rail System Using Analytic Hierarchy Process and Fuzzy Logic: A Case Study of Mumbai. *Urban Rail Transit*, 2019b, 5(3), 155–71.

Jasti, P.C., Ram, V.V. Sustainable benchmarking of a public transport system using analytic hierarchy process and fuzzy logic: A case study of Hyderabad, India. *Public Transp*, 2019a, 11(3), 457–85.

Jones, S., Tefe, M., Appiah-Opoku, S. Proposed framework for sustainability screening of urban transport projects in developing countries: A case study of Accra, Ghana. *Transp Res A: Policy Pract*, 2013, 49, 21–34.

Karolemeas, C., Tsigdinos, S., Tzouras, P.G., Nikitas, A., Bakogiannis, E. Determining Electric Vehicle Charging Station Location Suitability: A Qualitative Study of Greek Stakeholders Employing Thematic Analysis and Analytical Hierarchy Process. *Sustainability*, 2021, 13(4), 2298.

Keeney, R.L. Decision analysis: An overview. *Oper Res*, 30(5) (1982), 803–38.

Keeney, R.L., Raiffa, H. *Decision with Multiple Objectives*. Wiley, New York, 1976.

Kersuliene, V., Zavadskas, E.K., Turskis, Z. Selection of rational dispute resolution method by applying new step-wise weight assessment ratio analysis (Swara). *J Bus Econ Manag*, 2010, 11(2), 243–58.

Keshavarz-Ghorabaee, M., Zavadskas, E.K., Olfat, L., Turskis, Z. Multi-criteria inventory classification using a new method of evaluation based on distance from average solution (EDAS). *Informatica*, 2015, 26(3), 435–51.

Keshavarz-Ghorabaee, M.K., Zavadskas, E.K., Turskis, Z., Antucheviciene, J. A new combinative distance-based assessment (CODAS) method for multi-criteria decision-making. *Econ Comput Econ Cybern Stud Res*, 2016, 50(3), 25–41.

Kijewska, K., Torbacki, W., Iwan, S. Application of AHP and DEMATEL Methods in Choosing and Analysing the Measures for the Distribution of Goods in Szczecin Region. *Sustainability*, 2018, 10(7), 2365.

Kramar, U., Dragan, D., Topolsek, D. The Holistic Approach to Urban Mobility Planning with a Modified Focus Group, SWOT, and Fuzzy Analytical Hierarchical Process. *Sustainability*, 2019, 11(23), 6599.

Kumar, A., Anbanandam, R. Environmentally responsible freight transport service providers' assessment under data-driven information uncertainty. *J Enterp Inf Manag*, 2021, 34(1), 506–42.

Lee, J., Arts, J., Vanclay, F. Stakeholder views about Land Use and Transport Integration in a rapidly-growing megacity: Social outcomes and integrated planning issues in Seoul. *Sustain Cities Soc*, 2021, 67, 102759.

Ma, F., Shi, W.J., Yuen, K.F., Sun, Q.P., Guo, Y.R. Multi-stakeholders' assessment of bike sharing service quality based on DEMATEL-VIKOR method. *Int J Logist Res Appl*, 2019, 22(5), 449–72.

MacCrimon, K.R. *Decision Marking Among Multiple-Attribute Alternatives: A Survey and Consolidated Approach*. RAND memorandum, RM-4823-ARPA. The Rand Corporation, Santa Monica, 1968, 63.

Macharis, C., Milan, L. Transition through dialogue: A stakeholder based decision process for cities: The case of city distribution. *Habitat Int*, 2015, 45, 82–91.

Macharis, C., Turcksin, L., Lebeau, K. Multi actor multi criteria analysis (MAMCA) as a tool to support sustainable decisions: State of use. *Decis Support Syst*, 2012, 54(1), 610–20.

Mareschal, B., Brans, J.P. *PROMETHEE V: MCDM Problems with Segmentation Constrains.* Universite Libre de Brusells: Brussels, 1992; 13–30.

Mohagheghi, V., Mousavi, S.M., Aghamohagheghi, M., Vahdani, B. A new approach of multi-criteria analysis for the evaluation and selection of sustainable transport investment projects under uncertainty: A case study. *Int J Comput Intell Syst*, 2017, 10(1), 605–26.

Moslem, S., Campisi, T., Szmelter-Jarosz, A., Duleba, S., Nahiduzzaman, K.M., Tesoriere, G. Best-Worst Method for Modelling Mobility Choice after COVID-19: Evidence from Italy. *Sustainability*, 2020, 12(17), 6824.

Moslem, S., Ghorbanzadeh, O., Blaschke, T., Duleba, S. Analysing Stakeholder Consensus for a Sustainable Transport Development Decision by the Fuzzy AHP and Interval AHP. *Sustainability*, 2019, 11(12), 3271.

Munda, G. *Multi-criteria evaluation in a fuzzy environment.* Theory and applications in ecological economics Physica-Verlag, 1995, Heidelberg.

Munier, N. A new approach to the rank reversal phenomenon in MSDM with the SIMUS method. *Int J Multicriteria Decis Mak*, 2016, 11, 137–52.

Neofytou, H., Sarafidis, Y., Gkonis, N., Mirasgedis, S., Askounis, D. Energy Efficiency contribution to sustainable development: A multi-criteria approach in Greece. *Energy Sources B: Econ Plan Policy*, 2020, 15(10–12), 572–604.

Nowakowski, P., Krol, A. The influence of preliminary processing of end-of-life tires on transportation cost and vehicle exhausts emissions. *Environ Sci Pollut Res*, 2021, 28(19), 24256–69.

Ogrodnik, K. Multi-Criteria Analysis of Smart Cities in Poland. *Geographia Polonica*, 2020, 93(2), 163–81.

Oltean-Dumbrava, C., Miah, A. Assessment and relative sustainability of common types of roadside noise barriers. *J Clean Prod*, 2016, 135, 919–31.

Oltean-Dumbrava, C., Watts, G., Miah, A. Towards a more sustainable surface transport infrastructure: A case study of applying multi criteria analysis techniques to assess the sustainability of transport noise reducing devices. *J Clean Prod*, 2016, 112, 2922–34.

Oltean-Dumbrava, C., Watts, G., Miah, A. Transport infrastructure: Making more sustainable decisions for noise reduction. *J Clean Prod*, 2013, 42, 58–68.

Opricovic, S. *Multicriteria Optimization of Civil Engineering Systems.* PhD Thesis, Faculty of Civil Engineering, Belgrade, 1998, 302.

Ortega, J., Moslem, S., Toth, J., Peter, T., Palaguachi, J., Paguay, M. Using Best Worst Method for Sustainable Park and Ride Facility Location. *Sustainability*, 2020, 12(23), 10083.

Oses, U., Roji, E., Gurrutxaga, I., Larrauri, M. A multidisciplinary sustainability index to assess transport in urban areas: A case study of Donostia-San Sebastian, Spain. *J Environ Plan Manag*, 2017, 60(11), 1891–922.

Palevicius, V., Burinskiene, M., Podvezko, V., Paliulis, G.M., Sarkiene, E., Saparauskas, J. Research on the Demand for Parking Lots of Shopping Centres. *E M: Ekon Manag*, 2016, 19(3), 173–94.

Pamucar, D., Cirovic, G. The selection of transport and handling resources in logistics centers using Multi-Attributive Border Approximation Area Comparison (MABAC). *Expert Syst Appl*, 2015, 42(6), 3016–28.

Pamucar, D., Deveci, M., Canitez, F., Bozanic, D. A fuzzy Full Consistency Method-Dombi-Bonferroni model for prioritizing transportation demand management measures. *Appl Soft Comput*, 2020, 87, 105952.

Pamucar, D., Stevic, Z., Sremac, S. A New Model for Determining Weight Coefficients of Criteria in MCDM Models: Full Consistency Method (FUCOM). *Symmetry*, 2018, 10, 393.

Paul, A., Moktadir, M.A., Paul, S.K. An innovative decision-making framework for evaluating transportation service providers based on sustainable criteria. *Int J Prod Res*, 2020, 58(24), 7334–52.

Phdungsilp, A. Integrated energy and carbon modeling with a decision support system: Policy scenarios for low-carbon city development in Bangkok. *Energy Policy*, 2010, 38(9), 4808–17.

Pryn, M.R., Cornet, Y., Salling, K.B. Applying Sustainability Theory to Transport Infrastructure Assessment Using a Multiplicative AHP Decision Support Model. *Transport*, 2015, 30(3), 330–41.

Raiffa, H. *Decision analysis: Introductory lectures on choices under uncertainty (2nd print ed.)*, Volume Behavioral Science Quantitative Methods, Addison-Wesley, Reading, MA. 1970.

Rezaei, J. Best-worst multi-criteria decision-making method. *Omega*, 2015, 53, 49–57.

Roy, B. La methode ELECTRE. *Revue d'Informatique et. de Recherche Operationelle (RIRO)*, 1968, 8, 57–75.

Rybarczyk, G., Wu, C.S. Bicycle facility planning using GIS and multi-criteria decision analysis. *Appl Geogr*, 2010, 30(2), 282–93.

Rzesny-Cieplinska, J., Szmelter-Jarosz, A., Moslem, S. Priority-based stakeholders analysis in the view of sustainable city logistics: Evidence for Tricity, Poland. *Sustain Cities Soc*, 2021, 67, 102751.

Saaty, T.L. *Decision Making with Dependence and Feedback: The Analytic Network Process*. Pittsburg: RWS Publications, 1996, 34–72.

Saaty, T.L. *The Analytic Hierarchy Process*. McGraw-Hill: New York, 1980, 11–29.

Salabun, W., Palczewski, K., Watrobski, J. Multicriteria Approach to Sustainable Transport Evaluation under Incomplete Knowledge: Electric Bikes Case Study. *Sustainability*, 2019, 11(12), 3314.

Salabun, W.: The characteristic objects method: A new distance-based approach to multicriteria decision-making problems. *J MultiCriteria Decis Anal*, 2014, 22, 37–50.

Salling, K.B., Pryn, M.R. Sustainable transport project evaluation and decision support: Indicators and planning criteria for sustainable development. *Int J Sustain Dev World Ecol*, 2015, 22(4), 346–57.

San-Jose, J.T., Cuadrado, J. Industrial building design stage based on a system approach to their environmental sustainability. *Constr Build Mater*, 24(4) 2010, 438–47.

Sayyadi, R., Awasthi, A. An integrated approach based on system dynamics and ANP for evaluating sustainable transportation policies. *Int J Syst Sci: Oper Logist*, 2020, 7(2), 182–91.

Semanjski, I., Gautama, S. A Collaborative Stakeholder Decision-Making Approach for Sustainable Urban Logistics. *Sustainability*, 2019, 11(1), 234.

Shishegaran, A., Shishegaran, A., Mazzulla, G., Forciniti, C. A Novel Approach for a Sustainability Evaluation of Developing System Interchange: The Case Study of the Sheikhfazolah-Yadegar Interchange, Tehran. *Int J Environ Res Public Health*, 2020, 17(2), 435.

Shmelev, S.E., Shmeleva, I.A. Global urban sustainability assessment: A multidimensional approach. *Sustain Dev*, 2018, 26(6), 904–20.

Siksnelyte-Butkiene, I., Streimikiene, D. Sustainable Development of Road Transport in the EU: Multi-Criteria Analysis of Countries' Achievements. *Energies*, 2022, 15(21), 8291.

Simongati, G. Multi-Criteria Decision Making Support Tool for Freight Integrators: Selecting the Most Sustainable Alternative. *Transport*, 2010, 25(1), 89–97.

Soria-Lara, J.A., Banister, D. Evaluating the impacts of transport backcasting scenarios with multi-criteria analysis. *Transp Res Part A Policy Pract*, 2018, 110, 26–37.

Spadaro, I., Pirlone, F. Sustainable Urban Mobility Plan and Health Security. *Sustainability*, 2021, 13(8), 4403.

Stoilova, S., Munier, N., Kendra, M., Skrucany, T. Multi-Criteria Evaluation of Railway Network Performance in Countries of the TEN-T Orient-East Med Corridor. *Sustainability*, 2020, 12(4), 1482.

Tadic, S., Krstic, M., Roso, V., Brnjac, N. Dry Port Terminal Location Selection by Applying the Hybrid Grey MCDM Model. *Sustainability*, 2020, 12(17), 6983.

Tadic, S., Krstic, M., Roso, V., Brnjac, N. Planning an Intermodal Terminal for the Sustainable Transport Networks. *Sustainability*, 2019, 11(15), 4102.

Tang, J.Y., McNabola, A., Misstear, B. The potential impacts of different traffic management strategies on air pollution and public health for a more sustainable city: A modelling case study from Dublin. *Sustain Cities Soc*, 2020, 60, 102229.

Turskis, Z., Antucheviciene, J., Kersuliene, V., Gaidukas, G. Hybrid Group MCDM Model to Select the Most Effective Alternative of the Second Runway of the Airport. *Symmetry-Basel*, 2019, 11(6), 792.

Ullah, K., Hamid, S., Mirza, F.M., Shakoor, U. Prioritizing the gaseous alternatives for the road transport sector of Pakistan: A multi criteria decision making analysis. *Energy*, 2018, 165, 1072–84.

Vallee, D., Zielniewicz, P. ELECTRE III-IV. France, 1994; 18–38.

Van de Kaa, G., Scholten, D., Rezaei, J., Milchram, C. The Battle between Battery and Fuel Cell Powered Electric Vehicles: A BWM Approach. *Energies*, 2017, 10(11), 1707.

Vermote, L., Macharis, C., Putman, K. A Road Network for Freight Transport in Flanders: Multi-Actor Multi-Criteria Assessment of Alternative Ring Ways. *Sustainability*, 2013, 5(10), 4222–46.

Wang, C.N., Le, T.Q., Chang, K.H., Dang, T.T. Measuring Road Transport Sustainability Using MCDM-Based Entropy Objective Weighting Method. *Symmetry*, 2022, 14, 1033.

Wang, Z., Subramanian, N., Abdulrahman, M.D., Hong, C., Wu, L., Liu, C. Port sustainable services innovation: Ningbo port users' expectation. *Sustain Prod Consum*, 2017, 11, 58–67.

Watrobski, J., Malecki, K., Kijewska, K., Iwan, S., Karczmarczyk, A., Thompson, R.G. Multi-Criteria Analysis of Electric Vans for City Logistics. *Sustainability*, 2017, 9(8), 1453.

Web of Science database, 2022, available at: www.webofscience.com/wos/woscc/basic-search.

Yang, C.H., Lee, K.C., Chen, H.C. Incorporating carbon footprint with activity-based costing constraints into sustainable public transport infrastructure project decisions. *J Clean Prod*, 2016, 133, 1154–66.

Yazdani, M., Pamucar, D., Chatterjee, P., Chakraborty, S. Development of a decision support framework for sustainable freight transport system evaluation using rough numbers. *Int J Prod Res*, 2020, 58(14), 4325–51.

Yazdani, M., Zarate, P., Zavadskas, E.K., Turskis, Z. A combined compromise solution (CoCoSo) method for multi-criteria decision-making problems. *Manag Decis*, 2019, 57(9), 2501–19.

Zadeh L. A. Optimality and non-scalar-valued performance criteria. *IEEE Trans Automat Contr*, 1963, AC-8, 59–60.

Zagorskas, J., Turskis, Z. Location Preferences of New Pedestrian Bridges Based on Multi-Criteria Decision-Making and Gis-Based Estimation. *Balt J Road Bridge Eng*, 2020b, 15(2), 158–81.

Zagorskas, J., Turskis, Z. Setting Priority List for Construction Works of Bicycle Path Segments Based on Eckenrode Rating and Aras-F Decision Support Method Integrated in GIS. *Transport*, 2020a, 35(2), 179–92.

Zanakis, S.H., Solomon, A., Wishart, N., Dublish, S. Multi-attribute decision making: A simulation comparison of selected methods, *Eur J Oper Res*, 1998, 107(3), 507–29.

Zapolskyte, S., Vabuolyte, V., Burinskiene, M., Antucheviciene, J. Assessment of Sustainable Mobility by MCDM Methods in the Science and Technology Parks of Vilnius, Lithuania. *Sustainability*, 2020, 12(23), 9947.

Zavadskas, E.K., Kaklauskas, A., Sarka, V. The new method of multicriteria complex proportional assessment of projects. *Technol Econ Dev Econ*, 1994, 1(3), 131–39.

Zavadskas, E.K., Turskis, Z. A New Additive Ratio Assessment (ARAS) Method in Multicriteria Decision-Making. *Technol Econ Dev Econ*, 2010, 6(2), 159–72.

5 Multi-Criteria Decision Making for Sustainable Energy Development in Households

5.1 SUSTAINABLE ENERGY DEVELOPMENT IN HOUSEHOLDS

The questions of sustainable energy development are very important in various political documents in many developed and developing countries. The household sector consumes about 27 percent of the total energy generated and is the third largest energy consumer. Despite the fact, that greenhouse gas (GHG) emissions from households has a tendency to decrease in many regions of the world (Azizalrahman and Hasyimi, 2019), the household sector still causes 30–40 percent of total GHG emissions (Nejat et al., 2015). The results of the study by Ahlering et al. (2016) showed, that 93 percent of direct emissions from the residential sector are due to fuel combustion, especially for transport and heating needs. Therefore, it is recognized, that the development of renewable energy technologies (Yan et al., 2018), improvement in energy efficiency, and more sustainable energy usage are the key instruments to develop the energy sector in more sustainable ways and to mitigate climate change. However, despite technologies which have become cheaper in recent years, price is still one of the main barriers hindering new investments (Kiprop et al., 2019).

Different energy technologies have different pros and cons. However, if compared to fossil fuels, renewable energy sources (RES) provide a possibility to mitigate climate change and contribute to the protection of the environment. The development of renewable infrastructure is one of the most important aspects in transition to a low carbon economy. Therefore, the analysis and evaluation of different renewable energy technologies has been receiving huge attention in various countries' political documents and scientific literature. Fast development of various new and efficient renewable energy technologies in different regions of the world has been noted in recent years. Many countries have recognized that the usage of RES allows for moving towards a reliable, safe, and sustainable energy supply. At the local level, the use of RES not only combats the climate change, but also allows for improving the air quality, reduces energy dependency from imported energy sources, increases investments, promotes social and economic development of a country or a region by creating jobs and increases national or regional value, and so forth.

The costs of solar and wind energy technologies have significantly reduced in recent years, and their efficiency is quite high (IRENA, 2018). Therefore, these two

DOI: 10.1201/9781003327196-5

types of RES are the most popular in many countries. However, despite the existing achievements in renewable infrastructure and efficiency of technologies, it is very important to accelerate the development of new RES infrastructure in all regions of the world. The development of clean technologies in households is determined by many factors, such as: the economic situation, public awareness and acceptance, access to existing energy sources, policy measures implemented, and so forth.

As mentioned before, the improvements in energy efficiency is the other key aspect to the transition to low carbon energy society. About 40 percent of the total primary energy (Moran et al., 2017; Grygierek and Ferdyn-Grygierek, 2018) is consumed in the construction sector, which also emits about 10 percent of CO_2 emissions (Serghides et al., 2015). Therefore, the sector plays a significant role in addressing energy sustainability issues. Buildings renovation is a priority of EU Renovation Wave Strategy adopted in 2020 (European Commission, 2020). The strategy aims to double renovation rates until the end of 2030 and ensure improvements in energy efficiency and significant reduction of GHG emissions. The optimization of energy consumption in buildings is a very important aspect for the improvements in many climate change and energy related indicators (Santamouris, 2016). The biggest amount of energy is used to meet the heating, air conditioning and ventilation needs in buildings (Manzano-Agugliaro et al., 2015). Therefore, significant energy savings in buildings can be attained by selecting the optimal design solutions. For example, heat consumption can be effectively reduced with improvements in the insulation properties of buildings. The increase in buildings energy efficiency has become an important aspect of national energy strategies in many countries (Cao et al., 2016). There are many initiatives that focus on the construction sector with the objectives to promote technological innovation, improve energy efficiency (Noailly, 2012), reduce environmental impact (Goulden et al., 2017) and improve quality of life (Bonamente et al., 2018). The really big attention to the energy efficiency requirements is paid for the new buildings, but they account for only about 1 percent of the housing market annually (Serghides et al., 2015). It follows that old buildings must be renovated with a strong focus on energy efficiency requirements. For example, in the EU, the Energy Performance of Buildings Directive (EPBD) 2018/844 sets out particular requirements and objectives to be achieved (The European Parliament and the Council, 2018). The goal is that both new and renovated buildings become zero-energy buildings. It means the high energy efficiency and a big portion of RES in final energy consumption.

The really significant role for the energy efficiency of buildings is played building insulation materials. The selection of optimal thermal insulation materials is quite simple and one of the most popular strategies that can reduce the energy consumption in buildings effectively (Bisegna et al., 2016; Amani and Kiaee, 2020). The selection of insulation materials depends on many factors, such as thermal efficiency of the building, quality of life aspects, the impact on the environment (Aditya et al., 2017), and so forth. Currently, there is a wide range of insulation materials and each of them has different specific characteristics. Some insulation can be characterized as friendly environmentally, while others are more acceptable regarding economical criteria, and others have higher thermal characteristics (Al-Homoud, 2005; Patnaik et al., 2015; Asdrubali et al., 2015; Aditya et al., 2017; Gullbrekken et al., 2019), and so forth.

Also, it should be noted that the choice of materials in each case depends on many different factors, such as price, availability, transportation costs, climatic conditions, construction rules in the country, type of heating, and so forth. For example, in Europe, more than 60 percent of thermal insulation materials used are glass wool, stone wool, and inorganic fibrous materials. While the usage of organic foamy materials, polystyrene, expanded and extruded polystyrene does not reach 30 percent (Amani and Kiaee, 2020).

5.2 SELECTION OF INSTRUMENTS FOR THE RENEWABLE ENERGY TECHNOLOGIES SELECTION

This subsection introduces the most popular renewable energy technologies in the household sector and overviews previous studies, which were performed in order to compare different renewable energy technologies and/or hybrid energy systems, or were carried out for energy management decision making.

5.2.1 THE MAIN TYPES OF RENEWABLE ENERGY TECHNOLOGIES IN HOUSEHOLDS

The installation and usage of renewable energy technologies in households give many benefits. First of all, it improves the living conditions of household members by using energy more productively; helps to protect the quality of the environment; gives financial autonomy; contributes to sustainable planning and architecture (Jingchao and Kotani, 2012; Mahdavinejad and Karimi, 2012; World Bank, 2017; Kachapulula-Mudenda et al., 2018), and so forth. Energy sources could be divided into two main groups, which are fossil fuels and RES. Fossil fuels include natural gas, oil and coal, while renewable energy technologies include both conventional biomass and new sources such as solar, wind, and geothermal energy (Su et al., 2018). Solar photovoltaic (PV) and solar thermal, micro wind, heat pumps and small-scale biomass heating technologies can be distinguished as the most popular types of renewable energy technologies in households.

Solar PV and solar thermal technologies. It is known that solar energy can meet the total annual global energy demand, with an average of 1.6MWh/ m2 of energy per year (Li et al., 2020). However, annual solar radiation varies significantly in different regions of the world. Solar energy can be used to generate electricity and heat. Solar radiation is converted into thermal energy in solar collectors, and electricity is obtained directly from sunlight using PV cells. Thermal energy can be used for both: to heat homes and heat water. Solar power plants can operate independently or can be connected to the grid. In the case of autonomous operation, the energy produced is stored in accumulators, which ensure the energy supply in the event of demand. The energy storage system is not necessary if the power plant is connected to the power grid, and the electricity generated can be used not only for one's own consumption, but also the surplus electricity can be supplied to the grid. Solar energy is inexhaustible and the technological achievements in PV systems in the last decade have significantly increased the efficiency of technologies (Berrada and Loudiyi, 2016) and reduced installation costs (Li et al., 2020). In both solar PV and solar thermal technologies, the collectors in households are usually installed on rooftops or in other locations with direct sunlight.

Micro wind technology. It can be stated that wind energy is based on mature technologies and is developed by political incentives. Currently, increase in batteries' capacity and efficiency has boosted the development of new wind energy infrastructure (Berrada and Loudiyi, 2016; Zhang, 2018). Micro wind technology is a smaller device than conventionally used for wind energy generation and is suitable to fulfil private energy needs. Two types of wind turbines can be installed: vertical-axis or horizontal-axis wind turbines. The majority of households install devices on the rooftop or poles, and the efficiency of the technology depends on the characteristics of the device installed and the windiness in that specific area. The technology converts wind energy into electricity. The geographical characteristics of the location where the technology is installed are very important because generation is mainly based on the rotation speed of the wind turbine. Also, the turbines may be affected by potential obstructions nearby, such as trees and buildings, which can stop or turn the wind by preventing the turbines from operating at full capacity. Compared with solar energy technologies, the micro wind technology emits noise, which depends on the size and the device itself. Therefore, if the turbine is near the house or on the rooftop, it may affect quality of life conditions.

Heat pump technology. A heat pump can generate energy for heating, cooling, and preparation of hot water for residential, commercial, and industrial use. Heat pumps provide heating and cooling at the same time. Depending on the function a technology performs, it is called a heat pump, a cooling / refrigeration machine or an air-conditioning unit. Most of the energy generated is obtained from the environment: heat pumps can use renewable energy from air, ground, and water. Air source heat pumps use indoor, outdoor, or exhaust air as an energy source. Geothermal (or ground source) heat pumps use energy from the ground that is generated through a closed-loop vertical or horizontal collector. The energy obtained from the ground is transferred to brine or water and then to the device. Water source heat pumps work the same way as geothermal heat pumps. The only difference is that they use water directly instead of a closed-loop heat exchanger. Water heat pumps can be connected to lakes, rivers, sewage, cooling water, and so forth. (Chua et al., 2010). Also, the hybrid heat pump systems, as an example of a typical combination, can be an air source heat pump and small gas boiler, or a solar thermal collector and heat pump (Poppi et al., 2018), or a biomass boiler and heat pump. The efficiency of heat pumps depends on the efficiency of the devices themselves, the thermal energy needs of the building, and the climatic conditions (Sarbu and Sebarchievici, 2014).

Small-scale biomass heating technologies (pellet stoves and biomass boilers). Small-scale biomass technologies for heating are usually installed in private households. Wood and its by-products are the raw material for energy generation under this technology. Firewood, wood pellets, and wood chips are the most popular materials used to heat a private house. Firewood is the oldest and the most commonly used form of biomass. While, the popularity of wood chips has been growing rapidly only lately, because of the possibility to use them in automatic biomass heating systems. The selection of an automatic system has advantages for people who seek comfort and want to save time. The usage of wood chips is a better choice than the usage of firewood, because chips are made from wood waste, other wood products or directly from logs. But the best and most sustainable alternative is wood pellets. Compared to the alternatives, this type of fuel is the most sustainable choice and the most convenient. Wood pellets are made from sawdust and wood chips pressed under high pressure without any chemical additives. Wood pellets are distinguished by such criteria as being high in energy, easy to store and transport, and are optimal fuel solutions for small, fully automated heating systems. The new biomass boilers are distinguished for their high efficiency and low carbon monoxide (CO) emissions (Paniz, 2011).

Overall, it can be stated, that there are many contradictory aspects between different energy generation technologies. Each of them has different advantages and disadvantages compared with others and selection of the most appropriate depends on many aspects. For example, the most cost-effective does not always fit convenience criteria, nor is it the most environmentally friendly technology; the cheap and reliable energy supply does not always directly correlate with installation costs and payback, and so forth. Therefore, the application of MCDM techniques provide a possibility to assess these and other conflicting factors (Bhardwaj et al., 2019) and to find which alternative is the most suitable according to different criteria selected.

5.2.2   THE APPLICATION OF MCDM INSTRUMENTS

The application of MCDM techniques to solve questions in the energy sector is becoming more and more popular and now is one of the main methods for the assessment of different energy generation technologies. In the scientific literature can be found attempts to overview the application of MCDM tolls for the analysis and decision making of energy related issues. For example, an overview of MCDM approaches in scientific studies to solve energy sustainability issues are presented by Siksnelyte et al. (2018). Although the recognition that the development of innovative clean energy technologies in households could make a significant contribution to the development of sustainable energy systems and solve many energy-related problems (such as energy poverty, energy efficiency, energy security, etc.), there is a lack of studies that focus on the evaluation of different energy technologies in households. More active interest in such research can be observed only in the last few years.

The evaluation and analysis of different renewable energy technologies in households using MCDM techniques are carried out for different purposes in the scientific literature (Siksnelyte-Butkiene et al., 2020). The studies can be categorized as to the three main groups according to their objectives as follows:

- Technology comparison studies;
- Studies evaluating hybrid energy systems;
- Studies dealing with energy management issues.

Technology comparison studies. The first attempts to evaluate technologies can be characterized as having a quite limited amount of criteria. For example, the research carried out by Ren et al. (2009) introduced a methodology that can help to evaluate which energy generation technology is optimal for a household. The authors used linear programming and two MCDM techniques (AHP and PROMETHEE) for the calculations and ranking the alternatives selected. The presented methodology has been tested for a Japanese household. This study is one of the first attempts in the scientific literature to present a multi-criteria application for such kinds of research. According to the results, renewable energy technologies were not competitive with traditional energy generation technologies in the years under research. Also, it should be highlighted that only four criteria (two economic, one energy, one environmental) were selected for assessment and are very few if compared to the other research of this type. Ekholm et al. (2014) analyzed six different heating technologies for households. The authors measured the impact on health, climate, and acidification by the application of multi-criteria WSM techniques for each impact category separately. The results showed that none of the technologies is the best in all three impact categories. Nevertheless, the authors did not try to rank the selected heating technologies to find which one would be the best for households. It should be noted, that the criteria selected for this research consider only aspects related to pollution, without considering the economic or technological aspects that are very important for households.

Recent studies in the field have a greater inclusion of different criteria in the assessment. Also, the recent studies aim to determine the potential of technologies development. For example, an integrated evaluation model for buildings' energy supply systems was presented in the study by Dziugaite-Tumeniene et al. (2017). The model can support selecting the most optimal solution of energy supply for households. Also, the authors evaluated seven different renewable energy technologies, taking into account the household needs and characteristics of the building. The WASPAS method was applied for the calculations and ranking. The proposed evaluation model can also be applied to examine other energy generation technologies and their combinations. According to the results of the assessment, biomass and solar energy technologies have the biggest potential to reduce the impact on the environment of the energy system and to significantly increase the use of renewable energy. Yang et al. (2018) analyzed energy technologies and identified which of the alternatives should be developed in order to successfully implement the Danish energy strategy by 2035. One aim of this strategy is to replace oil boilers with renewables-based heating systems. The authors applied the multi-criteria TOPSIS technique as a primary tool to rank the alternatives. According to the criteria selected, the solar heating system is the most appropriate solution for private households, heat pumps are the second the most rational solution and wood pellet boilers took last place. Zhang et al. (2019) sought to create a multi-criteria evaluation system that allows for evaluation of the impact of public and private sectors on the usage of clean energy technologies in households. The developed system was applied for the Lithuanian case study. The application of three MCDM techniques (TOPSIS, EDAS, and WASPAS) showed that biomass boiler and solar thermal technologies are the most suitable solutions for Lithuanian households. Saleem and Ulfat (2019) applied the AHP method to assess renewable energy technologies. The authors performed an analysis of the Pakistani household electricity sector and analyzed the alternatives to meet household electricity demand. It was found that solar energy is the best renewable solution for Pakistani households in order to solve the issues related to a lack of electricity. Wind energy is the second best ranked solution, third is biomass, fourth is hydro energy, and last place in the ranking was ocean and geothermal energy. A new fuzzy integrated Delphi-AHP-PROMETHEE methodology was proposed by Seddiki and Bennadji (2019). The methodology can help to choose the best renewable energy technology to generate electricity. The proposed methodology includes the processes of expert interviews and responses analysis, criteria selection, and integration of multi-criteria techniques for the assessment.

Evaluation of hybrid energy systems. The other studies seek to evaluate hybrid energy systems, mainly for households in a certain location or region. For example, Jing et al. (2012) proposed a model for the assessment of combined cooling, heating and power system by integrating Fuzzy AHP approach. The presented model allows for evaluating energy technologies, taking into account economic, technical, social, and environmental aspects. The methodology was applied for the case study of one residential building in Beijing (China). Hacatoglu et al. (2015) sought to develop a methodology that can help to measure the sustainability of the energy system. The proposed methodology uses the WSM technique and applies economic, thermodynamic, and environmental criteria for sustainability assessment (Hacatoglu et al., 2013). On the basis of the proposed methodology, sustainability of hydrogen based storage with a solar-PV-wind-biomass energy system was assessed for 50 households in Ontario (Canada). According to the results, the proposed hybrid energy system was suitable solution to meet households needs. Vaisanen et al. (2016) applied the AHP approach and LCA to determine the most sustainable energy generation scenario in a small region in Norway. The three technological solutions for energy generation were evaluated, which were wind and hydro energy, hydro and solar power, and a small-scale CHP plant with solar electricity. According to the results of the assessment, the combined wind and hydro energy was the most sustainable solution under conditions of the study. The TOPSIS method and HOMER software were used in the study by Diemuodeke et al. (2019). The authors modelled hybrid energy alternatives and identified the best hybrid energy solutions in six regions of Nigeria, which had one of the world's largest electricity deficits in households. Babatunde et al. (2019) used two multi-criteria techniques (TOPSIS and CRITIC) and HOMER software to design the most suitable hybrid renewable energy system for low income households in a region of Nigeria.

Energy management issues. There is a particular lack of research that examines energy management issues for the development of renewable energy technologies in households. Although several research dealing with energy management issues can be mentioned. For example, Vasic (2018) evaluated six solutions for domestic heating in Serbia. The PROMETHEE method was applied for calculations and ranking the domestic heating alternatives. The study showed that special attention should be paid to the financial support mechanisms, because the economical aspects are the most important for those who invest in new renewable energy systems. Ferrer-Marti et al. (2018) developed a methodology for the analysis of household biogas digester programs and integrates the DEA-compromise programming approach. The methodology was applied for a case study in Latin America. The methodology created covers three levels of decision making, which are: community, digester model, and digester design selection. The DEA method was introduced by the Charnes method (Charnes et al.,1979), based on mathematical programming (Yang et al., 2015) and has been developed to evaluate efficiency. DEA measures the efficiency of a homogeneous set of decision making units based on their multiple inputs and outputs (Mardani et al., 2018; Yan et al., 2019). The technique is a valuable approach and is used alternatively or in addition to MCDM techniques.

Different renewable energy generation technologies have different advantages and disadvantages. However, if compared to fossil fuels, RES allow the solution of climate change and economic decarbonisation issues, which are so crucial today. Therefore, the analysis and assessment of clean energy technologies has been receiving increasing attention in both political documents of different countries and scientific literature. The household sector consumes almost one third of all energy generated; therefore tools and methodologies to solve different issues related to the energy consumption and development of clean energy in households are very important. Since the objectives from different perspectives (economic, environmental, social, technological) are very contradictory, the application of MCDM techniques allows for performing a comprehensive assessment and is one of the best ways to compare the alternatives.

5.3   CRITERIA FOR THE ASSESSMENT OF RENEWABLE ENERGY TECHNOLOGIES

Many different MCDM techniques can be applied to solve issues in the energy sector, and even more different criteria can be selected for the evaluation of technologies. Therefore, the aim of this subsection is to overview the scientific literature that applied MCDM techniques as a key instrument to evaluate renewable energy technologies in households. The commonly used criteria can be grouped in four categories, which are: economic, technological, environmental, and social. But the most commonly used were economical criteria. It was determined that the investment cost, operative, and maintenance cost are the most popular criteria in many studies analyzing different technologies in households (Table 5.1).

The most popular social criteria are sociocultural awareness / public acceptance, and criteria related to health. Also, the very significant role criteria play reflecting convenience issues of the technology used (Table 5.2).

The most commonly chosen technological criteria are related to daily comfort in using the technology, performance time of the equipment, renewable fraction, reliability, and efficiency of the technology and amount of energy generated. The summary of technological criteria is provided in Table 5.3

To evaluate the environmental impact of different household technologies the most popular way is to involve the criteria reflecting GHG intensity and impact on the environment: Water pollution, deforestation, and so forth (Table 5.4).

It is necessary to state that in the original studies, not all of the mentioned authors assigned the chosen evaluation criteria to these groups as presented above. For

TABLE 5.1
The Overview of Economic Criteria

Criteria	Technology comparison	Evaluation of hybrid energy systems	Energy management	Source
Initial capital cost / Investment cost / Affordability	+	+	+	Ren et al. (2009), Vaisanen et al. (2016); Dziugaite-Tumeniene et al. (2017); Yang et al. (2018); Vasic (2018);, Ferrer-Marti et al. (2018); Diemuodeke et al. (2019); Seddiki and Bennadji (2019); Saleem and Ulfat (2019)
Operative and maintenance cost / Annual costs of year	+	+	+	Ren et al. (2009); Dziugaite-Tumeniene et al. (2017); Ferrer-Marti et al. (2018); Vasic (2018); Saleem and Ulfat (2019); Babatunde et al. (2019); Seddiki and Bennadji (2019)
Total net present cost / Net present value	+	+		Dziugaite-Tumeniene et al. (2017); Babatunde et al. (2019); Diemuodeke et al. (2019); Seddiki and Bennadji (2019)
Return on investment / Payback period	+	+		Yang et al.(2018); Babatunde et al. (2019); Seddiki and Bennadji (2019)
Economic development / Commercial viability / Job creation	+	+	+	Hacatoglu et al. (2015); Vaisanen et al. (2016); Vasic (2018)
Reduced energy bill / Discount rate for year / Reduced operating expenses	+			Dziugaite-Tumeniene et al. (2017); Yang et al. (2018); Yang et al. (2018)
Cost of energy		+		Babatunde et al. (2019); Diemuodeke et al. (2019)
Production capacity	+			Saleem and Ulfat (2019)
Cost of fuel		+		Diemuodeke et al. (2019)
Resource availability		+		Hacatoglu et al. (2015)
Subsidy	+			Yang et al. (2018)
Residual value of technology	+			Dziugaite-Tumeniene et al. (2017)

Source: Created by authors.

TABLE 5.2
The Overview of Social Criteria

Criteria	Technology comparison	Evaluation of hybrid energy systems	Energy management	Source
Sociocultural awareness / Public acceptance	+	+	+	Vaisanen et al. (2016); Saleem and Ulfat (2019); Seddiki and Bennadji (2019); Ferrer-Marti et al. (2018); Babatunde et al. (2019); Diemuodeke et al. (2019)
Health / Footprint	+	+	+	Jing et al. (2012); Ekholm et al. (2014); Vaisanen et al. (2016); Ferrer-Marti et al. (2018)
Maintenance convenience	+	+	+	Jing et al. (2012); Ferrer-Marti et al. (2018); Seddiki and Bennadji (2019)
Affordability		+	+	Ferrer-Marti et al. (2018); Babatunde et al. (2019)
Land area used	+			Saleem and Ulfat (2019)
Job creation	+			Saleem and Ulfat (2019)
Number of potential beneficiaries			+	Ferrer-Marti et al. (2018)
Access to alternative fuels			+	Ferrer-Marti et al. (2018)
Local resources		+		Vaisanen et al. (2016)
Advanced performance		+		Jing et al. (2012)
Safeguards		+		Jing et al. (2012)

Source: Created by authors.

example, in the study carried out by Zhang et al. (2019), criteria were divided into two categories: cost criteria and benefit criteria. In addition to economic, technical, social, and environmental criteria, Vaisanen et al. (2016) included institutional criteria. These criteria reflect compatibility with Renewable Energy Directive, and consistency with the fiscal and public affairs policy of a country. For the evaluation of hybrid renewable energy systems, Hacatoglu et al. (2015) included thermodynamic criteria in addition to economic and environmental criteria. These criteria are very similar to technical criteria as they include energy efficiency and production capacity

TABLE 5.3
The Overview of Technological Criteria

Criteria	Technology comparison	Evaluation of hybrid energy systems	Energy management	Source
Comfort and easy to use	+		+	Jing et al. (2012); Yang et al.(2018); Vasic (2018); Ferrer-Marti et al. (2018)
Performance time of technology / Life cycle assessment	+	+	+	Ferrer-Marti et al. (2018); Saleem and Ulfat (2019); Diemuodeke et al. (2019)
Renewable fraction		+		Vaisanen et al. (2016); Babatunde et al. (2019); Diemuodeke et al. (2019)
Technology readiness and commercial maturity		+	+	Jing et al. (2012); Vasic (2018); Diemuodeke et al. (2019)
Efficiency	+			Yang et al.(2018); Saleem and Ulfat (2019); Seddiki and Bennadji (2019)
Reliability	+	+		Vaisanen et al. (2016); Yang et al. (2018); Seddiki and Bennadji (2019)
Energy production	+	+		Vaisanen et al. (2016); Babatunde et al. (2019); Saleem and Ulfat (2019)
Material availability	+		+	Ferrer-Marti et al. (2018); Seddiki and Bennadji (2019)
Unmet load		+		Babatunde et al. (2019); Diemuodeke et al. (2019)
Excess electricity		+		Babatunde et al. (2019)
Ease of installation		+		Babatunde et al. (2019); Diemuodeke et al. (2019)
Surface requirements			+	Ferrer-Marti et al. (2018)
Energy return on investment		+		Vaisanen et al. (2016)

(*continued*)

TABLE 5.3 (Continued)
The Overview of Technological Criteria

Criteria	Technology comparison	Evaluation of hybrid energy systems	Energy management	Source
Exergy efficiency		+		Jing et al. (2012)
Compatibility (grid electricity need)		+		Vaisanen et al. (2016)
Primary energy consumption ratio		+		Jing et al. (2012)
Regulation property		+		Jing et al. (2012)

Source: Created by authors.

TABLE 5.4
The Overview of Environmental Criteria

Criteria	Technology comparison	Evaluation of hybrid energy systems	Energy management	Source
GHG emissions / CO_2 emissions / SO_2 emissions	+	+	+	Ren et al. (2009); Jing et al. (2012); Ekholm et al. (2014); Hacatoglu et al. (2015); Vaisanen et al. (2016); Dziugaite-Tumeniene et al. (2017); Vasic (2018); Yang et al. (2018); Babatunde et al. (2019); Saleem and Ulfat (2019); Diemuodeke et al. (2019); Seddiki and Bennadji (2019)
Environmental impact	+	+	+	Hacatoglu et al. (2015); Vaisanen et al. (2016); Ferrer-Marti et al. (2018); Saleem and Ulfat (2019); Diemuodeke et al. (2019); Seddiki and Bennadji (2019)
Renewable fraction	+	+		Yang et al. (2018); Diemuodeke et al. (2019)

TABLE 5.4 (Continued)
The Overview of Environmental Criteria

Criteria	Technology comparison	Evaluation of hybrid energy systems	Energy management	Source
Agricultural land availability / land area		+	+	Vaisanen et al. (2016); Ferrer-Marti et al. (2018)
Natural resources availability		+		Babatunde et al. (2019)
Noise		+		Jing et al. (2012)

Source: created by authors.

factors. For the evaluation of heating technologies Ekholm et al. (2014) singled out the acidification criteria group in addition to environmental and social criteria. In addition to the four main groups of criteria, Seddiki and Bennadji (2019) singled out the usability and energetic criteria groups. However, the criteria involved in these two groups can complement the most popular four criteria groups. For example, energy production and ease of use can be classified as technical and easy to use or comfort level as social criteria. Especially close attention to the convenience and functionality of the technology was paid in the study by Dziugaite-Tumeniene et al. (2017). The authors presented the evaluation model that is oriented towards the user of technology and distinguished economic, energy, ecological, functionality, and convenience criteria. But despite the fact, that some of the criteria used by the different authors are divided into different groups, they can be grouped into the main four categories discussed above.

5.4 SELECTION OF INSTRUMENTS FOR THE BUILDING INSULATION MATERIALS SELECTION

The materials selection is not only one of the most important, but also one of the most challenging stages in building projection (Saghafi and Teshnizi, 2011). Many studies were performed with the aim to find the optimal, the most sustainable, or the most suitable materials for one project or another. The implementation of sustainable development goals (SDGs) can be achieved in different areas of economic activity. Therefore, not only technical characteristics and economic factors of materials are important for materials selection in the construction sector, but also social and environmental aspects should be taken into account (Samani et al., 2015). In order to reflect all contradictory aspects and the requirements of various stakeholders in the assessment, methodology its logical justification play a very significant role. The justified tools for evaluation and properly selected criteria allow for the solution of complicated questions and making reasonable decisions when choosing materials.

After careful literature analysis (Siksnelyte-Butkiene et al., 2021), the research studies evaluating insulation materials were grouped into three main categories:

- Sustainability assessment studies;
- Suitability assessment studies;
- Methods selection studies.

The recent studies evaluating insulation materials using MCDM methods belong to the *sustainability assessment category*. This shows the relevance of the topic. It is likely that research that will take into account sustainability aspects in the selection of insulation materials will only increase in the future. For example, an original framework for the sustainability assessment of insulation materials was created by Rocchi et al. (2018). The authors presented the case study from a farmhouse in central Italy, where the sustainability of 12 alternatives for roof insulation based on 7 criteria were considered. The criteria for the research were obtained by combining thermal comfort and energy optimization with the environmental and economic LLA and LCC analysis. The ELECTRE TRI-rC technique was applied for the organic and inorganic building insulation alternatives ranking. According to the results, the most suitable materials were polystyrene foam slab, kenaf fibers, hemp fibers, and cellulose. For the sustainability assessment of flat roof types, Guzman-Sanchez et al. (2018) created a set of 17 criteria, which reflect the SDGs of the United Nations. Two multi-criteria AHP and TOPSIS techniques were used for the calculations and ranking. The AHP method was used to determine the relative importance of criteria. The TOPSIS method was applied to rank the flat roof types. The assessment was performed taking into account the different possible weather scenarios. It was found that green roofs are the most sustainable option for all the scenarios created due to their insulation properties, recycling possibilities, embodied energy, life cycle cost, water purification and ecosystem-related characteristics. Rosasco and Perini (2019) sought to identify the factors that have the influence regarding the selection of building a roof system. The authors used the AHP approach for the evaluation of traditional and green roof systems. The experts were involved for the identification of the criteria and their weights. According to the experts, the most significant criteria were related to the category reflecting performance. The performed assessment indicated that a green roof is more preferable than the traditional roof. The interval TOPSIS technique was applied in the study by Streimikiene et al. (2020) for sustainability assessment and ranking of inorganic and organic building insulation materials. The authors performed the sensitivity analysis by creating four weighting scenarios (equal, balanced, environmental, technological). According to the results, the best solution for the three scenarios (equal, balanced, and technological) was recycled glass. Although sheep wool is the preferable option in the environmental scenario created.

Many different techniques have been applied for the *suitability assessment studies*. For example, Civic and Vucijak (2014) used the VIKOR method to evaluate eight building insulation materials. The authors chose seven criteria representing technical and environmental aspects. According to the results, the most suitable material was styrofoam, glass wool was in second place and wood wool was the third. The TOPSIS Grey technique was applied in the study by Zagorskas et al. (2014) to rank five modern insulation materials (eco wool, flax/hemp fiber, thermo wool, vacuum panel and aerogel) for historical buildings modernization. The results of the assessment showed that eco-wool was the best option. The design solutions for the residential buildings modernization by the application of the integrated SWARA-TODIM method was analyzed by Ruzgys et al. (2014). The authors carried out a case study in Lithuania and evaluated six alternatives of external wall insulation for building modernization (polystyrene foam and thin plaster; mineral wool, and fibre cement panels). The results showed that the most suitable alternative of residential building modernization is a ventilated system with 130 mm thickness mineral wool insulation and fibrocement panels. The innovative composite materials that incorporate rice husks and cork granules, were introduced in the study by Marques et al. (2020). The AHP technique was applied for calculations. According to the results of the experiment, the higher portion of rice husk in the composite formulations can lead to the better acoustic performance. While the expanded cork granules reduce the thermal conductivity. The four options of double-skin facades (corridor, multi-storey, box window, shaft-box) were assessed in the study by Bostancioglu (2020). The authors applied Fuzzy AHP technique to rank selected alternatives. The results showed that the box window was the best option according to the criteria selected, corridor facade was second, multi-storey double-skin facade was third, shaft-box took the last place in the ranking. The box window was the most suitable solution according to the three criteria (noise, fire protection, and thermal insulation). Finally, the authors compared the results of the study with previous research, where double-skin facades were assessed with AHP techniques (Bostancioglu and Onder, 2019). The results were confirmed, that is, unchanged. The WSM method was applied for the analysis of thermo modernizations solutions in the study by Basinska et al. (2020). The research was performed taking into account economic, energy-related and environmental criteria and more than 400 possible solutions were analyzed. It was found that the most preferable alternative is the solution of additional thermal insulation of extruded polystyrene with additional thicknesses of 30 cm and wood windows. According to the results, the use of insulation with a thickness above 36 cm does not result in a significant economic or energy effect.

A large variety of techniques and hybrid approaches can be found in studies proposing new *techniques for the materials* selection in the construction sector. For example, Ginevicius et al. (2008) used four MCDM techniques (SAW, TOPSIS, VIKOR, COPRAS) for ranking five external wall insulation alternatives for building renovation to find the most economically effective one. The methodology introduced by Zavadskas et al. (2008) allows for ranking different buildings' external wall design options. The presented methodology involves qualitative and quantitative criteria and is based on COPRAS technique. An approach to assess and rank different technologies in the construction sector can be found in the study by Zavadskas et al. (2013). The authors compared six mineral wool and polystyrene foam solutions for thermal insulation of external walls. The evaluation was performed on ELECTRE IV, MULTIMOORA and hybrid SWARA-TOPSIS, SWARA-VIKOR and SWARA-ELECTRE III approaches. In the other study by Zavadskas et al. (2017), the authors developed a tool for the materials selection in residential house construction. The tool is based on MULTIMOORA and Neutrosophic sets. The proposed new extension of MULTIMOORA was named as MULTIMOORA-SVNS. The study by Brauers et al. (2012) assess twenty roofs, external walls, ceilings, and windows solutions for the renovation of masonry buildings in Lithuania. The calculations and ranking were performed based on MOORA and MULTIMOORA techniques. A tool to rank different renovation alternatives was introduced by Seddiki et al. (2016). The proposed methodology is based on PROMETHEE technique with the combination of the Delphi method for criteria selection and the Swing method for the determination of criteria weights. The assessment presented a case study of a building in Algeria, where fifteen insulation solutions were compared. An approach for the decision making of building design facades was presented in the study by Moghtadernejad et al. (2020). The multi-criteria AHP technique and Choquet integrals are integrated in the approach proposed. The authors presented the guidelines for each design phase.

5.5 CRITERIA FOR THE ASSESSMENT OF BUILDING INSULATION MATERIALS

It is very popular to include the experts in the studies that evaluate building insulation materials. Usually, experts from the construction sector are involved. However, it can be found that studies also include the experts from science and employees of state authorities. The competencies of various experts can be used both for the selection of evaluation criteria and for the determination of criteria weights. Mainly, the studies which involve the experts in the evaluation process use their assistance to determine the weights of the criteria chosen. But not all the studies that used experts' assistance involved them in the process of criteria selection. An expert survey is the most popular applied way for criteria weighting. Usually, the importance of selected criteria is measured by pairwise comparison or by ranking from the most important to the least important. Other authors used their own estimation and expert surveys to determine weights (Ginevicius et al., 2008; Ruzgys et al. 2014), while the others applied Simon Roy Figueira's procedure (Rocchi et al., 2018), the Swing method (Seddiki et al., 2016). Surveys, Delphi method, and cross-group discussion (brainstorming technique) were used for the criteria selection.

Studies proposing methods for the selection of insulation materials in buildings also use the concordance coefficient by Kendal (Ginevicius et al., 2008; Zavadskas et al., 2008), criteria weighting by the application of SWARA method (Zavadskas et al., 2013; Ruzgys et al. 2014; Zavadskas et al., 2017). Table 5.5 presents techniques used in the scientific literature to select criteria and determine their weights in studies that used different MCDM methods for the insulation materials selection.

The criteria selected for the assessment of insulation materials are not categorized in most studies. Only a few authors divided criteria into several categories representing different dimensions of the assessment. For example, Seddiki et al. (2016) divided

TABLE 5.5
Criteria Selection and Determination of Their Weights

Source	MCDM method	Supporting instruments	Weighting	Experts, quantity	Type of experts	Criteria selection	Criteria groups
Zavadskas et al. (2008)	COPRAS, COPRAS-G	Kendal concordance coefficient	Experts survey (rating from the most important to the least important)	Yes, 39	Not specified	Own selection	Not specified
Ginevicius et al. (2008)	COPRAS, SAW, TOPSIS, VIKOR	Kendal concordance coefficient	Own estimation and expert survey (rating from the most important to the least important)	Yes, 16	Experts in construction (government, enterprises, researchers)	Experts survey	Not specified
Brauers et al. (2012)	MOORA, MULTIMOORA	-	N/A	No	-	Own selection	Not specified
Zavadskas et al. (2013)	ELECTRE III, ELECTRE IV, MULTIMOORA, TOPSIS, VIKOR	SWARA for weighting	Experts survey (rating from the most important to the least important), SWARA	Yes, 25	Experts in civil engineering and in heating, air conditioning and ventilation	Experts (Delphi method)	Not specified
Ruzgys et al. (2014)	TODIM	SWARA for weighting	Own estimation and expert survey (rating from the most to the least important)	Yes, 25	Not specified	Own selection	Not specified

(continued)

TABLE 5.5 (Continued)
Criteria Selection and Determination of Their Weights

Source	MCDM method	Supporting instruments	Weighting	Experts, quantity	Type of experts	Criteria selection	Criteria groups
Zagorskas et al. (2014)	TOPSIS	-	Pairwise comparison	Yes, 5	Experts in cultural heritage, climate change and energy	Own selection	Not specified
Civic and Vucijak (2014)	VIKOR	-	Own estimation	No	-	Own selection	Not specified
Seddiki et al. (2016)	PROMETHEE V	Sensitivity analysis	Swing method	Yes, 4	Experts in the construction and energy sector	Experts (Delphi method)	Economic, Energetic, Architectural
Zavadskas et al. (2017)	MULTIMOORA-SVNS	SWARA for weighting; sensitivity analysis; Neutrosophic sets	Pairwise comparison	Yes, 10	Experts in construction (architects, engineers and designer)	Own selection	Not specified criteria
Guzman-Sanchez et al. (2018)	AHP, TOPSIS	Sensitivity analysis	Experts (Questionnaire)	Yes, 50	Experts in the construction sector	Literature – the United Nations SDGs	Not specified

Study	Method		Weighting			Criteria derived from	Dimensions
Rocchi et al. (2018)	ELECTRE TRI-rC	Energy and comfort optimization, LCA, LCC, sensitivity analysis	Experts (Simon Roy Figueira procedure)	Not specified	Yes, 3	LCC and LCA analysis	Economic, Environmental
Bostancioglu and Onder (2019)	AHP	-	Experts (pairwise comparison, scale 1–9 (from 1 as "equally important" up to 9 as "extremely more important"))	Experts in the construction sector	Yes, 21	Literature	Not specified
Rosasco and Perini (2019)	AHP	-	Experts (pairwise comparison, scale 1–9 (from 1 as "equally important" up to 9 as "extremely more important"))	Experts in the construction sector (engineers, architects, researchers)	Yes, 30	Experts (cross-group discussion – brainstorming); Literature	Economic, Environmental, Social, Performance
Streimikiene et al. (2020)	interval TOPSIS	Sensitivity analysis	Own estimation (different weighting scenarios)	-	No	Own selection; Literature	Technological, Environmental
Bostancioglu (2020)	Fuzzy AHP	-	Experts (pairwise comparison, scale 1–9 (from 1 as "equally important" up to 9 as "extremely more important"))	Experts in the construction sector	Yes, 21	Literature	Not specified

(continued)

TABLE 5.5 (Continued)
Criteria Selection and Determination of Their Weights

Source	MCDM method	Supporting instruments	Weighting	Experts, quantity	Type of experts	Criteria selection	Criteria groups
Marques et al. (2020)	AHP	Sensitivity analysis	Own estimation (different weighting scenarios)	No	-	Own selection	Not specified
Moghtadernejad et al. (2020)	AHP	The Choquet integral	Experts (pairwise comparison, scale 1–9 (from 1 as "equally important" up to 9 as "extremely more important"))	No	-	Own selection; Literature	Not specified
Basinska et al. (2020)	WSM	LCA, Sensitivity analysis	The method presented by Mroz (2013)	No	-	Own selection; Literature	Not specified

Source: Created by authors.

the selected criteria into economic, energetic, and architectural for the assessment of facade renovation solutions. Rocchi et al. (2018) sought to measure the impact of sustainable insulations on the environment and economic suitability. The authors grouped the criteria into economic and environmental categories. Rosasco and Perini (2019) identified economic, performance, social, and environmental criteria for the identification of factors that have the biggest influence on the selection of roof systems. For the sustainability assessment of organic and inorganic insulation materials, Streimikiene et al. (2020) identified technological and environmental criteria categories.

The growing attention to the sustainability issues in the construction sector requires reconsidering criteria used for the selection materials. Although the criteria used in the scientific research are not divided into groups, this can be done in order to identify the popularity of the applied criteria and representation of different sustainability aspects. The criteria used in the previous studies are divided into four categories, which represent different dimensions of sustainable development (Table 5.6).

Basically, all the studies use indicators reflecting technological aspects. The most popular is to include criteria characterized thermal insulation properties. The application of the water absorption coefficient and installation complexity/duration of works are the second most popular criteria. Also, many studies take into account such criteria as durability, weight, fire classification, and noise insulation. Also, almost all studies use economic criteria. Mostly, the economic dimension is reflected by the investment costs. As the second most popular criteria reflecting economical aspects, energy losses or energy savings can be singled out. Indicators representing the environmental aspects are included in approximately half of the studies. The most commonly used indicators were CO_2 emissions/global warming indicators and environmental friendliness/resource sustainability of materials. Indicators reflecting social aspects can be found in almost half the studies, where the most popular is the aesthetic impact of materials used.

The results of the literature review allowed for identifying a variety of MCDM methods application for the selection of insulation materials in buildings. The applied methods have different characteristics and different possibilities, including data in the calculations. Table 5.7 provides the main advantages and disadvantages of MCDM techniques that were applied in previous studies for the assessment of insulation materials in buildings.

The most commonly used methods can be reviewed in more detail. The most popular technique applied is AHP, which was developed by Saaty (1980). The technique allows for solving multi-criteria questions using a pairwise comparison scale. The calculation process is quite simple and computation results are obtained relatively quickly compared to other multi-criteria methods. Also, the method can be easily applied in different fields, such as construction, energy, and so forth. (Kaya et al., 2018). The method is rational and based on a hierarchical structure, and therefore focuses on all criteria selected. However, it is necessary to highlight that the experience of decision makers plays a very important role, when determining the weights of the criteria. This aspect can complicate the assessment process if there is

TABLE 5.6
Overview of Criteria

Dimension	Criteria	Source
Technological	Thermal insulation properties	Ginevicius et al., 2008; Zavadskas et al., 2008; Brauers et al., 2012; Zavadskas et al., 2013; Zagorskas et al., 2014; Civic and Vucijak, 2014; Zavadskas et al., 2017; Guzman-Sanchez et al., 2018; Rosasco and Perini, 2019; Bostancioglu and Onder, 2019; Bostancioglu, 2020; Streimikiene et al., 2020; Moghtadernejad et al., 2020; Marques et al., 2020
	Water absorption	Ginevicius et al., 2008; Zavadskas et al., 2013; Zagorskas et al., 2014; Ruzgys et al. 2014; Civic and Vucijak, 2014; Streimikiene et al., 2020; Moghtadernejad et al., 2020; Marques et al., 2020
	Duration of works	Ginevicius et al., 2008; Zavadskas et al., 2008; Zavadskas et al., 2013; Zagorskas et al., 2014; Ruzgys et al. 2014; Bostancioglu and Onder, 2019; Bostancioglu, 2020; Moghtadernejad et al., 2020
	Durability / Risk of the material	Ginevicius et al., 2008; Zavadskas et al., 2008; Seddiki et al., 2016; Zavadskas et al., 2017; Rosasco and Perini, 2019; Moghtadernejad et al., 2020
	Weight	Ginevicius et al., 2008; Zavadskas et al., 2008; Zavadskas et al., 2017; Guzman-Sanchez et al., 2018; Rosasco and Perini, 2019; Moghtadernejad et al., 2020
	Noise insulation	Guzman-Sanchez et al., 2018; Rosasco and Perini, 2019; Bostancioglu and Onder, 2019; Bostancioglu, 2020; Moghtadernejad et al., 2020; Marques et al., 2020
	Fire classification	Bostancioglu and Onder, 2019; Bostancioglu, 2020; Streimikiene et al., 2020; Moghtadernejad et al., 2020
	Loss of space / Thickness	Zagorskas et al., 2014; Moghtadernejad et al., 2020
	Density	Civic and Vucijak, 2014; Streimikiene et al., 2020
	Specific heat	Civic and Vucijak, 2014; Streimikiene et al., 2020
	Wind pressure resistance	Bostancioglu and Onder, 2019; Bostancioglu, 2020;
	Daylight	Bostancioglu and Onder, 2019; Bostancioglu, 2020
	Warranty period	Ginevicius et al., 2008
	Protection	Guzman-Sanchez et al., 2018
	Adhesive joint strength	Ginevicius et al., 2008

TABLE 5.6　(Continued)
Overview of Criteria

Dimension	Criteria	Source
	Extraction force of a pin fixing thermal insulating board to solid materials	Ginevicius et al., 2008
	Wall load-bearing capacity	Zavadskas et al., 2017
Economic	Investment cost / Price	Ginevicius et al., 2008; Zavadskas et al., 2008; Brauers et al., 2012; Zavadskas et al., 2013; Zagorskas et al., 2014; Ruzgys et al. 2014; Civic and Vucijak, 2014; Seddiki et al., 2016; Zavadskas et al., 2017; Rosasco and Perini, 2019; Bostancioglu and Onder, 2019; Bostancioglu, 2020; Moghtadernejad et al., 2020
	Energy losses / Energy savings	Brauers et al., 2012; Ruzgys et al. 2014; Seddiki et al., 2016; Rocchi et al., 2018; Rosasco and Perini, 2019
	Payback period	Brauers et al., 2012; Zavadskas et al., 2013; Ruzgys et al. 2014
	Operation, maintenance and disposal costs	Rosasco and Perini, 2019; Moghtadernejad et al., 2020
	Annual energy consumption / Primary energy index	Moghtadernejad et al., 2020; Basinska et al., 2020
	Tax incentives	Rosasco and Perini, 2019
	Life cycle cost	Guzman-Sanchez et al., 2018
	Real estate benefit	Rosasco and Perini, 2019
	Net present value	Rocchi et al., 2018
	Global cost	Basinska et al., 2020
Environmental	CO_2 emissions / Global warming	Civic and Vucijak, 2014; Guzman-Sanchez et al., 2018; Rocchi et al., 2018; Rosasco and Perini, 2019; Streimikiene et al., 2020; Basinska et al., 2020
	Resource sustainability / Environmental friendliness	Zavadskas et al., 2017; Guzman-Sanchez et al., 2018; Rosasco and Perini, 2019; Moghtadernejad et al., 2020
	Solar power / Window solar performance	Guzman-Sanchez et al., 2018; Moghtadernejad et al., 2020
	Biodiversity	Guzman-Sanchez et al., 2018; Rosasco and Perini, 2019

(continued)

TABLE 5.6 (Continued)
Overview of Criteria

Dimension	Criteria	Source
	Non-renewable energy	Rocchi et al., 2018
	Runoff attenuation	Guzman-Sanchez et al., 2018
	Albedo coefficient	Guzman-Sanchez et al., 2018
	Agricultural productivity	Guzman-Sanchez et al., 2018
	Reduction of runoff temperature	Guzman-Sanchez et al., 2018
	Embodied carbon	Guzman-Sanchez et al., 2018
	Embodied energy	Guzman-Sanchez et al., 2018
	Water purification	Guzman-Sanchez et al., 2018
Social	Aesthetic	Seddiki et al., 2016; Zavadskas et al., 2017; Guzman-Sanchez et al., 2018; Rosasco and Perini, 2019; Bostancioglu and Onder, 2019; Bostancioglu, 2020; Moghtadernejad et al., 2020
	Health / Respiratory inorganics	Rocchi et al., 2018; Rosasco and Perini, 2019; Streimikiene et al., 2020
	Air quality and heat island reduction	Rosasco and Perini, 2019
	Comfort performance	Rocchi et al., 2018

Source: Created by authors.

more than one decision maker. Also, additional analysis is needed to verify the results of the assessment (Saaty, 2004; Ishizaka and Labib, 2009; Shahroodi et al., 2012; Kumar et al., 2017).

The TOPSIS technique is the second most applied method in such research. As mentioned in previous chapters, the method presented by Hwang and Yoon (1981) is based on measuring the distance to the ideal solution (Jato-Espino et al., 2014). As the AHP technique, the TOPSIS is distinguished for its fairly easy calculation process and quickly obtained results; the logic of computation is rational and clear, expressed in a quite simple mathematical form. Therefore, it is easy for the decision maker to interpret the results obtained and to understand the importance of the criteria selected for the final ranking. But it is necessary to highlight that TOPSIS is based on the Euclidean distance, which means that positive and negative values of criteria are not reflected in the computations. Also, it is necessary to mention, that the significant deviation from the best solution in one evaluation criterion has a significant impact on the final results (Shih et al., 2007; Boran et al., 2009). Therefore, technique is not proper tool when the criteria differ significantly among themselves.

TABLE 5.7
Comparison of MCDM Techniques

MCDM techniques	AHP	TOPSIS	MULTIMOORA	COPRAS	ELECTRE	VIKOR	MOORA	PROMETHEE	SAW	SWARA	TODIM	WSM
Popularity to assess building insulation materials	Very high	Very high	High	Ordinary	Ordinary	Ordinary	Rare	Rare	Rare	Rare	Rare	Rare
Advantages — Easy to computation process	x	x	x			x	x		x			x
Non-compensatory			x		x	x	x	x				
Intelligible logic of calculations		x	x			x	x					
Robust to outliers						x		x				
Disadvantages — Additional analysis is required to verification of the results	x				x				x			x
The application requires subjective assumptions	x				x			x	x	x		

Source: Created by authors.

The MOORA method, which was introduced by Brauers (Brauers, 2004), is recognized as an objective instrument. The technique is based on the ratio system and the reference point techniques, where desirable and undesirable criteria are used simultaneously for ranking (Brauers and Zavadskas, 2006). The method is widely applied due to its comprehensible logic of calculation and simplicity. Also, it is recognized as more robust than many other multi-criteria techniques. Brauers and Zavadskas (2010) added the full multiplicative form to the MOORA and the new technique was named MULTIMOORA. Accordingly, MULTIMOORA lies on three approaches, which are as follows: the ratio system and the reference point techniques, and the full multiplicative form (Zavadskas et al., 2015). Both methods are widely used to solve problems in different fields.

Zavadskas et al. presented the COPRAS method in 1994 (Zavadskas et al., 1994). COPRAS is one of the compromise techniques, where the ratio to the best ideal solution and the ratio to the worst ideal solution are determined. The method uses stepwise ranking and assessment procedure in terms of significance and degree of utility. Also, it is worth mentioning that both qualitative and quantitative data can be used in calculations.

The first version of the ELECTRE technique was presented by Roy in 1968 (Roy, 1968). The method requires determination of the concordance and discordance indices that lengthen the calculation process. Also, the subjective human intervention is required, because a decision maker has to select threshold values for the calculation of concordance and discordance indices (Karande and Chakraborty, 2012). It is recognized that additional analysis is required to verify results of the assessment.

The VIKOR method was introduced by Opricovic (1998) in 1998 and is widely used in different fields to make the most optimal decisions. It is worth mentioning, that it is quite popular to integrate VIKOR with other multi-criteria techniques (Mardani et al., 2016). The method measures the closeness to the positive and the negative ideal solutions. Different from the TOPSIS technique, the VIKOR takes into account the relative importance of the distance from the ideal solutions (Opricovic and Tzeng, 2007). It is scientifically recognized that the VIKOR approach is understandable and processing the calculation is quite simple. Despite that advantage, it should be mentioned that the results of the assessment can be affected by the normalization procedure and weighting strategy.

However, not only these, but also many other MCDM techniques can be used to select the best solutions in the construction sector. Depending on the experience of the decision maker, the available information, the possible time cost, and other factors, a variety of methods can be applied for different decision making problems.

REFERENCES

Aditya, L., Mahlia, T.M.I., Rismanchi, B., Ng, H.M., Hasan, M.H., Metselaar, H.S.C., Muraza, O., Aditiya, H.B. A review on insulation materials for energy conservation in buildings. *Renew Sust Energ Rev*, 2017, 73, 1352–65.

Ahlering, M., Fargione, J., Parton, W. Potential carbon dioxide emission reductions from avoided grassland conversion in the northern Great Plains. *Ecosphere*, 2016, 7 (12), e01625.

Al-Homoud, M.S. Performance characteristics and practical applications of common building thermal insulation materials. *Build Environ*, 2005, 40, 353–66.

Amani, N., Kiaee, E. Developing a two-criteria framework to rank thermal insulation materials in nearly zero energy buildings using multi-objective optimization approach. *J Clean Prod*, 2020, 276, 122592.

Asdrubali, F., D'Alessandro, F., Schiavoni, S. A review of unconventional sustainable building insulation materials. *Sustain Mater Technol*, 2015, 4, 1–17.

Azizalrahman, H., Hasyimi, V. A model for urban sector drivers of carbon emissions. *Sustain Cities Soc*, 2019, 44, 46–55.

Babatunde, O.M., Munda, J.L., Hamam, Y. Selection of a Hybrid Renewable Energy Systems for a Low-Income Household. *Sustainability*, 2019, 11(16), 4282.

Basinska, M., Kaczorek, D., Koczyk, H. Building Thermo-Modernisation Solution Based on the Multi-Objective Optimisation Method. *Energies*, 2020, 13(6), 1433.

Berrada, A., Loudiyi, K. Operation, sizing, and economic evaluation of storage for solar and wind power plants. *Renew Sustain Energy Rev*, 59, 2016, 1117–29.

Bhardwaj, A., Joshi, M., Khosla, R., Dubash, N.K. More priorities, more problems? Decision-making with multiple energy, development and climate objectives. *Energy Res Soc Sci*, 2019, 49, 143–57.

Bisegna, F., Mattoni, B., Gori, P., Asdrubali, F., Guattari, C., Evangelisti, L., Sambuco, S., Bianchi, F. Influence of Insulating Materials on Green Building Rating System Results. *Energies*, 2016, 9, 712.

Bonamente, E., Brunelli C., Castellani, F. Garinei, A., Biondi, L., Marconi, M., Piccioni, E. A life-cycle approach for multi-objective optimisation in building design: Methodology and application to a case study. *Civ Eng Environ Syst*, 2018, 35(1–4), 158–79.

Boran, F.E., Genc, S., Kurt, M., Akay, D. A multi-criteria intuitionistic fuzzy group decision making for supplier selection with TOPSIS method. *Expert Syst Appl*, 2009, 36, 11363–68.

Bostancioglu, E. Double skin facade assessment by fuzzy AHP and comparison with AHP. *Archit Eng Des Manag*. 2020.

Bostancioglu, E., Onder, N.P. (2019) Applying analytic hierarchy process to the evaluation of double skin façades. *Archit Eng Des Manag*, 2019, 15(1), 66–82.

Brauers, W.K. *Optimization Methods for a Stakeholder Society. A Revolution in Economic Thinking by Multiobjective Optimization*. Boston: Kluwer Academic Publishers, 2004, p. 352.

Brauers, W.K.M., Kracka, M., Zavadskas, E.K. Lithuanian Case Study of Masonry Buildings from the Soviet Period. *J Civ Eng Manag*, 2012, 18(3), 444–56.

Brauers, W.K.M., Zavadskas, E.K. Project Management by MULTIMOORA as an Instrument for Transition Economies. *Technol Econ Dev Econ*, 2010, 16(1), 5–24.

Brauers, W.K.M., Zavadskas, E.K. The MOORA method and its application to privatization in transition economy. *Control Cybern*, 2006, 35(2), 443–68.

Cao, X., Dai, X., Liu, J. Building energy-consumption status worldwide and the state-of-the-art technologies for zero-energy buildings during the past decade. *Energy and Build*, 2016, 128, 198–213.

Charnes, A., Cooper, W.W., Rhodes E. Measuring the efficiency of decision making units. *Eur J Oper Res*, 1979, 2(6), 429–44.

Chua, K.J., Chou, S.K., Yang, W.M. Advances in heat pump systems: A review. *Appl Energy*, 87(12), 2010, 3611–24.

Civic, A., Vucijak, B. Multi-criteria Optimization of Insulation Options for Warmth of Buildings to Increase Energy Efficiency. *Procedia Eng*, 2014, 69, 911–20.

Diemuodeke E.O., Addo A., Oko, C.O.C., Mulugetta, Y., Ojapah M.M. (2019) Optimal mapping of hybrid renewable energy systems for locations using multi-criteria decision-making algorithm. *Renew Energy*, 134, 461–77.

Dziugaite-Tumeniene, R., Motuziene, V., Siupsinskas, G., Ciuprinskas, K., Rogoza, A. Integrated assessment of energy supply system of an energy-efficient house. *Energy and Build*, 2017, 138, 443–54.

Ekholm, T., Karvosenoja, N., Tissari, J., Sokka, L., Kupiainen, K., Sippula, O., Savolahti, M., Jokiniemi, J., Savolainen, I. A multi-criteria analysis of climate, health and acidification impacts due to greenhouse gases and air pollution-The case of household-level heating technologies. *Energy Policy*, 2014, 74, 499–509.

European Commission. A Renovation Wave for Europe – greening our buildings, creating jobs, improving lives. Communication from the Commission to the European Parliament, the Council, the European Economic and Social Committee and the Committee of the Region. Brussels, 2020, COM (2020) 662 final.

Ferrer-Marti L., Ferrer, I., Sanchez, E., Garfi, M. A multi-criteria decision support tool for the assessment of household biogas digester programmes in rural areas. A case study in Peru. *Renewable Sustainable Energy Rev*, 2018, 95, 74–83.

Ginevicius, R., Podvezko, V., Raslanas, S. Evaluating the Alternative Solutions of Wall Insulation by Multicriteria Methods. *J Civ Eng Manag*, 2008, 14(4), 217–26.

Goulden, S., Erell, E., Garb, Y., Pearlmutter, D. Green building standards as socio-technical actors in municipal environmental policy, *Build Res Inf*, 2017, 45(4), 414–25.

Grygierek, K., Ferdyn-Grygierek, J. Multi-Objective Optimization of the Envelope of Building with Natural Ventilation. *Energies*, 2018, 11, 1383.

Gullbrekken, L., Grynning, S., Gaarder, J.E. Thermal Performance of Insulated Constructions – Experimental Studies. *Buildings*, 2019, 9, 49.

Guzman-Sanchez, S., Jato-Espino, D., Lombillo, I., Diaz-Sarachaga, J.M. Assessment of the contributions of different flat roof types to achieving sustainable development. *Build Environ*, 2018, 141, 182–92.

Hacatoglu, K., Dincer, I., Rosen, M. A. Sustainability assessment of a hybrid energy system with hydrogen-based storage. *Int J Hydrog Energy*, 2015, 40(3), 1559–68.

Hacatoglu, K., Dincer, I., Rosen, M.A. A Methodology to Assess the Sustainability of Energy Systems through Life-Cycle Analysis and Sustainability Indicators. *The 3rd World Sustainability Forum*, November 1–30, 2013, p. 20.

Hwang, C.L., Yoon, K. *Multiple Attributes Decision Making Methods and Applications*. Springer: Berlin, Heidelberg, 1981, 22–51.

IRENA. Global energy transformation: A roadmap to 2050, *International Renewable Energy Agency*, 2018, Abu Dhabi.

Ishizaka, A, Labib, A. Analytic hierarchy process and expert choice: Benefits and limitations. *OR Insight*, 2009, 22(4), 201–20.

Jato-Espino, D., Castillo-Lopez, E., Rodriguez-Hernandez, J., Canteras-Jordana, J.C. A review of application of multi-criteria decision making methods in construction. *Autom Constr*, 2014, 45, 151–62.

Jing, Y.Y., Bai, H., Wang, J.J. A fuzzy multi-criteria decision-making model for CCHP systems driven by different energy sources. *Energy Policy*, 2012, 42, 286–96.

Jingchao, Z., Kotani, K. The determinants of household energy demand in rural Beijing: Can environmentally friendly technologies be effective? *Energy Econ*, 2012, 34(2), 381–8.

Kachapulula-Mudenda, P., Makashini, L., Malama, A., Abanda, H. Review of Renewable Energy Technologies in Zambian Households: Capacities and Barriers Affecting Successful Deployment. *Building*, 2018, 8(6), 77.

Karande, P., Chakraborty, S. Application of multi-objective optimization on the basis of ratio analysis (MOORA) method for materials selection. *Mater Des*, 2012, 37, 317–24.

Kaya, I., Çolak, M., Terzi, F. Use of MCDM techniques for energy policy and decision-making problems: A review. *Int J Energy Res*, 2018, 42, 2344–72.

Kiprop, E., Matsui, K., Maundu, N. The Role of Household Consumers in Adopting Renewable Energy Technologies in Kenya. *Environments*, 2019, 6(8), 95.

Kumar, A., Sah, B., Singh, A. R., Deng, Y., He, X., Kumar, P., Bansal, R.C. A review of multi criteria decision making (MCDM) towards sustainable renewable energy development. *Renew Sust Energ Rev*, 2017, 69, 596–609.

Li, H.X., Edwards, D.J., Hosseini, M.R., Costin, G.P. A review on renewable energy transition in Australia: An updated depiction. *J Clean Prod*, 2020, 242, UNSP 118475.

Mahdavinejad, M., Karimi, M. Challenges and Opportunities Regarding Adoption of Clean Energy Technology in Developing Countries, in Case of Iran. *Int J Smart Grid Clean Energy*, 2012, 2, 283–88.

Manzano-Agugliaro, F., Montoya, F.G., Sabio-Ortega, A., García-Cruz, A. Review of bioclimatic architecture strategies for achieving thermal comfort. *Renew Sustain Energy Rev.* 2015, 49, 736–55.

Mardani, A., Streimikiene, D., Balezentis, T., Saman, M.Z.M., Nor, K.M., Khoshnava, S.M. Data Envelopment Analysis in Energy and Environmental Economics: An Overview of the State-of-the-Art and Recent Development Trends. *Energies*, 2018, 11(8), 2002.

Mardani, A., Zavadskas, E.K., Govindan, K., Senin, A.A., Jusoh, A. VIKOR Technique: A Systematic Review of the State of the Art Literature on Methodologies and Applications. *Sustainability*, 2016, 8, 37.

Marques, B., Tadeu, A., Antonio, J., Almeida, J., de Brito, J. Mechanical, thermal and acoustic behaviour of polymer-based composite materials produced with rice husk and expanded cork by-products. *Constr Build Mater*, 2020, 239, 117851.

Moghtadernejad, S., Chouinard, L.E., Mirza, M.S. Design strategies using multi-criteria decision-making tools to enhance the performance of building facades. *J Build Eng*, 2020, 30, 101274.

Moran, P., Goggins, J., Hajdukiewicz, M. Super-insulate or use renewable technology? Life cycle cost, energy and global warming potential analysis of nearly zero energy buildings (NZEB) in a temperate oceanic climate. *Energy Build.* 2017, 139, 590–607.

Mroz, T.M. *Energy Management in Built Environment: Tools and Evaluation Procedures*. Poznan University of Technology: Poznan, Poland, 2013, p. 138. ISBN 8377752387.

Nejat, P., Jomehzadeh, F., Taheri, M.M., Gohari, M., Muhd, M.Z. A global review of energy consumption, CO2 emissions and policy in the residential sector (with an overview of the top ten CO2 emitting countries). *Renew Sustain Energy Rev*, 2015, 43, 843–62.

Noailly, J. Improving the energy efficiency of buildings: The impact of environmental policy on technological innovation. *Energy Econ*, 2012, 34(3), 795–806.

Opricovic, S. *Multicriteria Optimization of Civil Engineering Systems*. PhD Thesis, Faculty of Civil Engineering, Belgrade, 1998, p. 302.

Opricovic, S., Tzeng, G.H. Extended VIKOR method in comparison with outranking methods. *Eur J Oper Res*, 2007, 178, 514–29.

Paniz, A. Working Group 2 on Small scale heating systems. Handbook. AIEL – Italian Agriforestry Energy Association, CrossBorder Bioenergy: IEE/09/933/SI2.558306, 2011, p. 22.

Patnaik, A., Mvubu, M., Muniyasamy, S., Botha, A., Anandjiwala, R.D. Thermal and sound insulation materials from waste wool and recycled polyester fibers and their biodegradation studies. *Energy and Build*, 2015, 92, 161–69.

Poppi, S. Sommerfeldt, N., Bales, C., Madani, H., Lundqvist, P. Techno-economic review of solar heat pump systems for residential heating applications. *Renew Sust Energ Rev*, 81(1), 2018, 22–32.

Ren, H., Gao, W., Zhou, W., Nakagami, K. Multi-criteria evaluation for the optimal adoption of distributed residential energy systems in Japan. *Energy Policy*, 2009, 37(12), 5484–93.

Rocchi, L., Kadzinski, M., Menconi, M.E., Grohmann, D., Miebs, G., Paolotti, L., Boggia, A. Sustainability evaluation of retrofitting solutions for rural buildings through life cycle approach and multi-criteria analysis. *Energy and Build*, 2018, 173, 281–90.

Rosasco, P., Perini, K. Selection of (Green) Roof Systems: A Sustainability-Based Multi-Criteria Analysis. *Buildings*, 2019, 9, 134.

Roy, B. La methode ELECTRE. Revue d'Informatique et. de Recherche Operationelle (RIRO), 1968, 8, 57–75.

Ruzgys, A., Volvaciovas, R., Ignatavicius, C., Turskis, Z. Integrated evaluation of external wall insulation in residential buildings using SWARA-TODIM MCDM method. *J Civil Eng Manage*, 2014, 20(1), 103–10.

Saaty, T.L. Decision making – the analytic hierarchy and network processes (AHP/ANP). *J Syst Sci Syst*, 2004, 13(1), 1–35.

Saaty, T.L. *The Analytic Hierarchy Process*. McGraw-Hill: New York, 1980, 11–29.

Saghafi, M.D., Teshnizi, Z.S.H. Recycling value of building materials in building assessment systems. *Energy and Build*, 2011, 43, 3181–88.

Saleem, L., Ulfat, I. A Multi Criteria Approach to Rank Renewable Energy Technologies for Domestic Sector Electricity Demand of Pakistan. *Mehran Univ Res J Eng Technol*, 2019, 38(2), 443–52.

Samani, P., Mendes, A., Leal, V., Guedes, J.M., Correia, N. A sustainability assessment of advanced materials for novel housing solutions. *Build Environ*, 2015, 92, 182–91.

Santamouris, M. Innovating to zero the building sector in Europe: Minimizing the energy consumption, eradication of the energy poverty and mitigating the local climate change. *Sol Energy*, 2016, 128, 61–94.

Sarbu, I., Sebarchievici, C. General review of ground-source heat pump systems for heating and cooling of buildings. *Energy and Build*, 70, 2014, 441–54.

Seddiki, M., Anouche, K., Bennadji, A., Boateng, P. A multi-criteria group decision-making method for the thermal renovation of masonry buildings: The case of Algeria. *Energy and Build*, 2016, 129, 471–83.

Seddiki, M., Bennadji, A. Multi-criteria evaluation of renewable energy alternatives for electricity generation in a residential building. *Renew Sust Energ Rev*, 2019, 110, 101–17.

Serghides, D.K., Dimitriou, S., Katafygiotou, M.C., Michaelidou, M. Energy efficient refurbishment towards nearly zero energy houses, for the Mediterranean region. *Energy Procedia*, 2015, 83, 533–43.

Shahroodi, K., Keramatpanah, A., Amini, S., Sayyad Haghighi, K. Application of analytical hierarchy process (AHP) technique to evaluate and selecting suppliers in an effective supply chain. *Kuwait Chapter Arab J Bus Manag Rev*, 2012, 1(6), 119–32.

Shih, H.S., Shyur, H.J., Lee, E.S. An extension of TOPSIS for group decision making. *Math Comput Model*, 2007, 45, 801–13.

Siksnelyte, I., Zavadskas, E.K., Streimikiene, D., Sharma, D. An Overview of Multi-Criteria Decision-Making Methods in Dealing with Sustainable Energy Development Issues. *Energies*, 2018, 11(10), 2754.

Siksnelyte-Butkiene, I., Streimikiene, D., Balezentis, T., Skulskis, V. A Systematic Literature Review of Multi-Criteria Decision-Making Methods for Sustainable Selection of Insulation Materials in Buildings. *Sustainability*, 2021, 13(2), 737.

Siksnelyte-Butkiene, I., Zavadskas, E.K., Streimikiene, D. Multi-Criteria Decision-Making (MCDM) for the Assessment of Renewable Energy Technologies in a Household: A Review. Energies, 2020, 13, 1164.

Streimikiene, D., Skulskis, V. Balezentis, T., Agnusdei, G.P. Uncertain multi-criteria sustainability assessment of green building insulation materials. *Energy and Build*, 2020, 219, 110021.

Su, W., Liu, M., Zeng, S., Streimikiene, D., Balezentis, T., Alisauskaite-Seskiene, I. Valuating renewable microgeneration technologies in Lithuanian households: A study on willingness to pay. *J Clean Prod*, 2018, 191, 318–29.

The European Parliament and the Council. Directive 2018/844 of the European Parliament and the Council of the 30 May 2018 Amending Directive 2010/31/EU on the Energy Performance of Buildings and Directive 2012/27/EU on Energy Efficiency. *O J Eur Union*, 2018, 156, 75–91.

Vaisanen, S., Mikkila, M., Havukainen, J., Sokka, L., Luoranen, M., Horttanainen, M. Using a multi-method approach for decision-making about a sustainable local distributed energy system: A case study from Finland. *J Clean Prod*, 2016, 137, 1330–38.

Vasic, G. Application of multi criteria analysis in the design of energy policy: Space and water heating in households – City Novi Sad, Serbia. *Energy Policy*, 2018, 113, 410–19.

World Bank. *State of Electricity Access Report*; World Bank: Washington, DC, 2017.

Yan, Q.Y., Zhao, F., Wang, X., Yang, G.L., Balezentis, T., Streimikiene, D. The network data envelopment analysis models for non-homogenous decision making units based on the sun network structure. *Cent Eur J Oper Res*, 2019, 27(4), 1221–44.

Yang, L., Ouyang, H., Fang, K.N., Ye, L.L., Zhang, J. Evaluation of regional environmental efficiencies in China based on super-efficiency-DEA. *Ecol Indic*, 2015, 51, 13–19.

Yang, Y., Ren, J., Solgaard, H.S., Xu, D., Nguyen, T.T. Using multi-criteria analysis to prioritize renewable energy home heating technologies. *Sustain Energy Technol Assess*, 2018, 29, 36–43.

Zagorskas, J., Zavadskas, E.K., Turskis, Z., Burinskiene, M., Blumberga, A., Blumberga, D. Thermal insulation alternatives of historic brick buildings in Baltic Sea Region. *Energy and Build*, 2014, 78, 35–42.

Zavadskas, E.K., Antucheviciene, J., Hajiagha, S.H.R., Hashemi, S.S. The Interval-Valued Intuitionistic Fuzzy MULTIMOORA Method for Group Decision Making in Engineering. *Math Probl Eng*, 2015, 560690.

Zavadskas, E.K., Bausys, R., Juodagalviene, B., Garnyte-Sapranaviciene, I. Model for residential house element and material selection by neutrosophic MULTIMOORA method. *Eng Appl Artif Intell*, 2017, 64, 315–24.

Zavadskas, E.K., Kaklauskas, A., Sarka, V. The new method of multicriteria complex proportional assessment of projects. *Technol Econ Dev Econ*, 1994, 1(3), 131–39.

Zavadskas, E.K., Kaklauskas, A., Turskis, Z., Tamosaitiene, J. Selection of the effective dwelling house walls by applying attributes values determined at intervals. *J Civ Eng Manag*, 2008, 14(2), 85–93.

Zavadskas, E.K., Turskis, Z., Volvaciovas, R., Kildiene, S. Multi-criteria Assessment Model of Technologies. *Stud Inform Control*, 2013, 22(4), 249–58.

Zhang, C.H., Wang, Q., Zeng, S.Z., Balezentis, T., Streimikiene D., Alisauskaite-Seskiene, I., Chen, X.L. Probabilistic multi-criteria assessment of renewable micro-generation technologies in households. *J Clean Prod*, 2019, 212, 582–92.

Zhang, X., Li, H.-Y., Deng, Z.D., Ringler, C., Gao, Y., Hejazi, M.I., Leung, L.R. Impacts of climate change, policy and Water-Energy-Food nexus on hydropower development. *Renew Energy*, 2018, 116, 827–34.

Conclusions

(1) Sustainable development goals (SDGs) cover the most important economic, social, environmental, and technological targets that allow for moving toward a better future and welfare of all nations. It is not possible to arrive at a final fixed state of sustainability for any country. It can be assumed that each generation will have more understanding and more advanced technologies and other innovations and will have different priorities based on their needs linked to their cultures and values so the path toward sustainability is almost infinite. Consequently, SDGs must be consistently reorganized. Scholars agree that, in order to implement SDGs, it is necessary to find a balance between conflicting interests of various stakeholders at different levels over time that necessitates a multi-disciplinary approach in decision making.

(2) Energy plays a crucial role in achieving sustainable development as it is the main driver of economic development and the major source of environmental pollution, including GHG emissions. Therefore, questions of sustainable energy development are of huge importance in decision making on all levels, including local, national, regional, and global. Sustainable energy development aims to ensure production and use of energy resources in ways that are compatible with long-term human well-being and ecological balance. Therefore, for developing policies to promote sustainable energy development, it is necessary to trade off between sometimes conflicting economic, social, and environmental targets.

(3) Different MCDM approaches have been successfully applied to solving problems of sustainable energy development, and the usage of different techniques has begun to be more and more popular in recent years. MCDM methods allow for assessing alternatives by taking into account many criteria. The rationality, effectiveness, and scientific recognition of the methods allow for considering which are the most suitable instruments for dealing with issues related to the main problems of sustainable energy development.

(4) The presented comparative evaluation of the most popular MCDM techniques highlights the importance of the nature of the data, the experience of the decision maker, the accuracy of the results, and possible time costs.

DOI: 10.1201/9781003327196-6

The highlighted characteristics of the MCDM methods show alternatives that allow for speeding up the selection of decision making instruments for future research.

(5) Sustainability in the energy sector is a fundamental principle in shaping European energy policy and in solving urgent challenges facing EU member states. The most important issues for EU energy policy have been the same for many years: energy dependency on non-EU suppliers, security of energy supply, climate change and protection of the environment, and so forth. Now, in the face of the COVID-19 pandemic and global economic uncertainty due to the global energy crisis and war in Ukraine, the EU government has many more challenges, one of the largest being a sharp increase in energy prices. The developed framework for the assessment of implementation of energy policy priorities considers the most critical sustainable energy development indicators. The practical example of presented framework was applied for the case study of BSR countries. To reflect the different situations of countries and the ability to achieve energy policy objectives, economic, environmental, and social indicators were selected following the traditional concept of sustainability.

(6) The framework for the electricity sector sustainability assessment was developed and applied to measure and rank the achievements of the EU countries. A framework of indicators was created to reflect economic, environmental, and security of supply issues of power sector development. The performed analysis concludes that electricity markets can be characterized by low rivalry and passive participation of consumers in most EU member states. The share of renewables in gross final power consumption is very low in many EU countries. Therefore, it is critical to use policies and measures actively to encourage a smarter and more sustainable electricity generation. It can be concluded that despite various innovations like smart grids, smart meters, and smart homes, customers are still not enough informed and stimulated to participate actively in the energy market or to become energy prosumers. Consequently, customers lose the capability to manage their electricity consumption, which has direct impact on savings cost and energy efficiency.

(7) The framework for a heating sector sustainability assessment is based on a traditional concept of sustainability and considers the main economic, environmental, and social indicators reflecting the sector. The developed framework was applied for the case study of North European countries. The results showed that the most sustainable heating sectors are in the Nordic countries. Nordic countries were ranked first in all weighing scenarios created. The main conditions for the development of sustainable heating sectors can be singled out as the major share of RES in a final energy mix and a significant portion of district heating. The developed framework can be employed universally to track the progress achieved, to measure the sustainability of other countries or regions, and so forth. However, the geographical differences among countries should be considered before application of the framework.

(8) The developed road transport sustainability assessment framework was applied for the measurement of EU member states' progress made in approaching sustainable transport during the last decade. The framework accounts for the main indicators revealing the main issues relevant to sustainability of road transport and is a suitable instrument for sustainability evaluation and analysis of achievements made. Also, the presented framework can be straightforwardly used to monitor the progress achieved in the future. The results of such assessment can serve for the identification of the best performing EU member states in order to track the best practices implemented. The analysis can be performed, not only for EU countries' ranking and comparison, but also can be adapted for other countries or regions.

(9) The detailed content analysis of studies dealing with sustainable transport development by applying different MCDM tools and the categorization by the application areas, revealed that in most cases the multi-criteria analysis has been focused on sustainable transport planning and sustainability assessment issues. The performed analysis presents the main insights for methods selection for future research. Also, the thematic areas for criteria selection that reflect sustainable development concepts are provided.

(10) The sustainable energy development targeting the household sector is a central core, because the household sector is one of the main contributors to GHG emissions. The conducted analysis of studies using MCDM methods for the evaluation of renewable energy technologies in households and assessment of insulation materials in buildings allows for understanding the potential of MCDM techniques for the decision making process. Provided frameworks of indicators will make easier the design of future studies in this field. The evaluation of advantages and disadvantages of MCDM techniques allows the defining of the most suitable methods and tools for assessment and ranking of sustainable energy options in the household sector.

(11) The evaluation of renewable energy microgeneration technologies in households by applying MCDM techniques was carried out for different purposes. According to the main goal of the studies, the researches can be grouped into three main categories, such as: comparison of technologies; solution of energy management problems, and assessment of hybrid energy systems. Also, after careful analysis, the most commonly used criteria for technologies comparison can be singled out as economic, technological, social, and environmental. Investment, operation, and maintenance costs were the most popular economic criteria in analyzed studies. The main technological criteria found in scientific studies were the share of RES, technology, performance time, and reliability. The main social criteria were found to be sociocultural awareness and public acceptance. For the assessment of environmental issues linked to various microgeneration technologies, the most popular criterion was GHG intensity.

(12) The scientific literature analysis dealing with insulation materials in buildings using MCDM instruments allowed for singling out the main three groups of studies: sustainability assessment, suitability assessment, and

method selection studies. The content analysis showed that the most popular multi-criteria method used for comparison of insulation materials is AHP. The second most commonly applied technique is TOPSIS. Both methods can be characterized as having a quite simple computation process and can be easy applied in practice. Also, they do not require high time costs or specific knowledge of the decision maker. Although studies in the methods selection category proposed more multifaceted algorithms, the instruments were much more methodologically sound, with the most attention being paid to criteria selection and criteria weighting.

(13) In most studies evaluating building insulation materials, the criteria selected for comparison of materials are not grouped into categories. However, it can be stated that all the studies used indicators that reflect technological properties. The criteria indicating thermal insulation characteristics are the most popular. Also, it is popular to involve economic aspects in the assessment. Mostly, the economic dimension was reflected by investment costs. The criteria measuring social or environmental aspects were reflected in approximately one half of the studies. However, in the context of global energy crisis, and in order to perform a comprehensive comparison of different insulation materials, criteria used for the assessment should represent all sustainability dimensions. The performed analysis of the criteria and their grouping helps other scholars in the selection of sustainability assessment criteria and their justification for further studies in this area.

Index

A

Agricultural, 6, 145
  productivity, 156
Air pollution, 3, 54
  source, 111
Affordability, 2–3, 8–9, 11, 15–20, 54–55, 81, 90, 92–93, 111, 141
Action plan, 12, 51, 75
Atmosphere, 15, 51, 54–55

B

Baltic States, 49, 64, 87
Bicycle, 118
Bioenergy, 6, 86, 89
Biogas, 6, 139
Biomass, 61, 65, 86–88, 135–136, 138
  boiler, 136–138
  conversation, 86
  energy, 139
    system, 139
  heating, 136
    systems, 136
  sources, 88
  technologies, 136

C

Case study, 48, 117–120, 123, 139–140, 146–148, 166
Citizens, 26, 121, 123
Clean energy, 2, 6, 14, 21, 51, 75, 140
  measures, 27
  sources, 6
  technologies, 21, 137–138, 140
Climate change, 2, 8–10, 12, 15–16, 29, 45, 47–49, 51, 54, 62, 65–68, 73–74, 76, 78, 89, 94, 111, 133–134, 140, 150, 166
  goals, 46, 123
  mitigation, 8, 46, 89
  objectives, 65–66
  policy, 89, 91
  problems, 11
  targets, 75
CO_2 emissions, 7, 46, 134, 144, 155
  standards, 90
Coal, 7, 48–49, 76, 78, 83, 86–87, 135
  reserves, 48
Comparative assessment, 67, 74, 78, 90
Consumer behavior, 20

Cooling, 46, 74–76, 136, 139
  decarbonization, 73
  needs, 73
  sector, 46, 75–76, 80
  systems, 73, 80
  targets, 45
  water, 136
Cooperation, 10, 21, 64, 120
COVID-19, 47
  lockdowns, 47
  pandemic, 94, 117, 166
Cumulative market, 55

D

Decarbonization, 46–47, 51–52, 65, 67, 73–74, 78, 81, 88–90
Decision makers, 25–27, 29, 35, 38, 54, 111, 122, 153
Deforestation, 140
Demand management, 49
Density, 154
Dependency, 8, 16–17, 19, 48–49, 55, 64, 66, 68, 72–74, 166
  indicators, 50
  rate, 49–50
Developed countries, 3, 20–21, 45
Developing countries, 3–4, 17–18, 20–21
Disparities, 93, 97
Distribution losses, 55, 68
Distribution, 3, 16, 19, 68, 72, 93
  activities, 68
  losses, 55, 68
  services, 121
Durability, 153–154

E

Economic development, 2, 11, 13, 141
  levels, 17
Economic efficiency, 19
Economic expansion, 1
Economic growth, 1–3, 17
Ecosystems, 2–3, 6
Electric vehicles, 73, 90, 123
Electricity, 2, 45, 47, 55–56, 61, 64, 67, 71–73, 82–83, 86–87, 90, 135–136, 139, 143–144
  consumption, 79
  coridors, 48
  demand, 72–73, 139

distribution, 72
generation, 77–78, 87, 90, 135
infrastructure, 73
interconnection, 63, 68
market, 66–67
network, 49, 63–64, 72
policy, 67
prices, 67–68
ring, 64
sector, 46, 51, 55, 64–65, 67–70, 73–74, 90, 139
system, 64
Electrification, 66, 73, 90
Energy accessibility, 11, 54
Energy activism, 16
Energy affordability, 8–9, 16, 18
measures, 16
Energy demand, 45, 47, 49, 69
Energy dependency, 48, 64, 66, 166
Energy efficiency, 8, 9–11, 18–20, 46, 51, 55–56,
62, 64–65, 74–76, 79, 81–82, 86, 123, 134,
137, 142, 166
awareness, 18
directive, 51, 74–75
improvements, 8–9, 18–19, 51
barriers, 20
issues, 74, 86
level, 81
measures, 16, 65, 123
objectives, 62
requirements, 134
savings, 81
standards, 21
targets, 49, 51, 64–65, 75
technologies, 20
Energy equity, 4, 6, 16, 18
Energy grids, 19
Energy import, 16, 19
Energy independence, 55, 65, 76
Energy infrastructure, 29, 49, 64–65, 69
projects, 64
Energy initiatives, 10, 25, 27
Energy intensity, 55
Energy justice, 16
concept, 16
issues, 18
Energy market, 3, 19, 21, 48–49, 63–64, 67, 166
Energy mix, 48, 62–63, 65, 68, 74–75, 77–78,
87–88, 166
Energy needs, 18, 46, 50, 69, 136
Energy policy, 12, 26, 33, 37, 48, 61, 65, 67, 69,
74, 82, 88, 166
achievements, 66, 68
aspects, 29, 47
challenges, 48
context, 66, 74, 89

goals, 47, 49, 60, 73
issues, 32
objectives, 19, 60, 74, 115, 166
priorities, 48, 51, 56, 65–66, 166
problems, 66
scenarios, 123
targets, 65, 73
Energy poverty, 10, 16, 18, 80, 82–83
assessment, 35
issues, 76
studies, 81
Energy price, 11, 16, 29, 48, 54, 61, 64, 69, 166
Energy producers, 20–21
Energy production, 2, 4, 6, 10, 16–17, 19–20, 54,
143, 145
Energy productivity, 54–55
Energy saving, 16
ratio, 79
Energy security, 2, 4, 6, 9, 11, 16–18, 48–50, 56,
64, 67–69, 74, 83, 137, 166
issues, 6, 11–12, 48, 66, 68, 76
level, 63
problems, 74
strategy, 49
Energy sources, 2–3, 6, 15–18, 49, 62, 65, 75–76,
135
diversification, 49
Energy supply, 2–4, 6, 8–9, 12, 16–19, 21, 49, 54,
63–65, 76, 79–80, 133, 135, 137–138, 166
affordability, 2
crisis, 49
curves, 12
problems, 49
security, 2, 8
issues, 2
services, 6
systems, 18, 138
Energy sustainability, 4, 29, 73, 112
issues, 70, 134, 137
problems, 116
Energy system, 2, 4, 8–10, 12–13, 15–20, 45, 51,
54, 64–65, 67, 73, 80, 82, 86–87, 135,
137–139, 142–145
Energy transition, 7, 11, 18, 29, 45–46, 51, 88–89,
94
policy, 29
Energy vulnerability, 16
Engagement, 26–27, 112, 119–121, 123–124
Environmental sustainability, 4, 6, 80
Environmentalist, 1
Equity, 2, 4, 10, 16, 17–18, 54–55, 98
issues, 14
EU energy policy, 47–48, 51, 69, 166
achievements, 66
goals, 60

objectives, 47
    priorities, 48, 56, 65–66
        optimization, 56, 114
European Economic Area, 49
European Environment Agency, 11, 54, 67
European Green Deal, 7, 89

F

Final energy, 46
    balance, 75
    consumption, 51, 52, 55, 61, 64, 66, 68, 74–75, 134
    mix, 62, 65, 78, 166
Financial institutions, 20
Firewood, 136–137
Footprint, 79, 142
Fossil fuels, 2–3, 7–8, 12, 15, 18, 46, 61, 64–66, 75–79, 83, 87–88, 90, 133, 135, 140
    account, 66
    combustion, 51
    power plants, 7
Fuel oil, 7

G

Gas pipelines, 48
GDP, 2, 55, 77–79
    growth, 2
GHG emission, 7, 9, 18–19, 45–47, 51, 55–56, 62–63, 65–66, 74–75, 77–78, 81–83, 87–89, 91, 94–97, 111, 123–124, 133–134, 144, 165, 167
    emitters, 46
    from transport, 47
    in Europe, 51
    in the EU, 47
    level, 90, 94
    per capita, 77, 87
    per GDP, 77
    per person, 94
    reduction, 6, 8, 10, 19, 52, 89, 91
        avoidance, 18
        scenarios, 12
        targets, 10
    target, 64, 66
Global warming, 89, 153, 155
Globalization, 2–3, 21
Governments, 12, 19–21
Green New Deal, 7
Green recovery, 7
Gro Harlem Brundtland, 1

H

Heating sector, 46, 64, 74, 76, 78, 81–83, 86–88
    sustainability, 74, 78–81, 83–84

assessment, 74, 82, 166
    issues, 81
Holistic, 9, 47
    approach, 9, 20, 123
    way, 47
Households, 18, 68, 71, 78, 80–81, 85, 133–140, 167
    needs, 139
    sector, 19, 55
Human health, 2–3, 15, 83, 88, 111
Hybrid energy, 135, 137, 139, 167
    alternatives, 139
    solutions, 139
    systems, 135, 137, 139, 167
Hybrid vehicles, 94
Hydrogen vehicles, 90

I

Import dependency, 16, 49
    indicators, 50
    rate, 49–50
Income poverty, 16
Independency, 55, 72
Indicators, 7, 30–34, 36–37, 39, 46–52, 54–57, 60, 62, 67–69, 77–84, 86–89, 91–93, 95, 111, 120, 122, 134, 153
    for sustainable development, 7, 13, 17
    set, 33, 35, 67
    selection, 35, 112, 124
    weights, 37
    for sustainability assessment, 47, 91
    system, 120
Inequality, 12
Installation, 3, 18, 90, 135, 153
    costs, 135, 137, 143
Interconnection, 49, 63–64, 68, 72
International Atomic Energy Agency, 7, 11, 54, 67
International Energy Agency, 11, 54, 67
International energy, 11, 17, 54, 67
    agency, 11, 54, 67
    economy, 17
        issues, 17
International policy, 2, 11, 14
Investment cost, 79, 81, 140–141, 155, 168
Investments, 3–4, 18, 20–21, 25, 46, 54, 65, 75, 78–79, 81, 90, 97, 122, 133

K

Knowledge, 25–27, 29, 39, 112, 118, 124, 168
Kyoto protocol, 9

L

Labelling, 51
    schemes, 20, 80
Liberalization, 3, 19, 67

Life Cycle Assessment, 79, 120
Lignite, 7
Low-carbon energy, 7, 11, 18, 21, 66, 89, 134
    sector, 21, 66
    society, 134
    transition, 7, 11, 18, 89
Low-income, 78

M

Millennium Development Goals, 10
Mobility, 91, 95, 97, 111, 117, 119–120
    alternatives, 115
    choices, 117
    system, 118
Modernization, 4, 64–65, 82, 147
Multi-criteria, 30, 32, 37, 70, 80, 83, 121, 146
    AHP techniques, 148
    analysis, 29, 31–32, 37, 47, 112, 115, 124, 167
        application, 47
        techniques, 70, 120, 122, 139
    applications, 90, 138
    ASPID technique, 80
    assessment, 59
        results, 71, 84, 95
    BWM technique, 117
    decision analysis, 112
    decision making, 30, 45, 56, 111, 112, 115, 133
    evaluation, 83
        system, 138
    methods, 30, 119, 153, 168
    questions, 153
    techniques, 29, 32, 158
    TOPSIS technique, 138
    WSM technique, 138
Multidimensional concept, 6
Multi-objective, 56
    optimization, 56, 114

N

National energy, 65, 87
    consumption, 65
    efficiency, 64
    independence, 65
    need, 49
    sources, 49, 65
    strategies, 134
    system, 45, 80
Natural gas, 7, 48–49, 62, 64, 76, 83, 85, 86–88,
    135
Nuclear, 6, 64
    energy, 29
    fuels, 17
    power, 6
    waste, 6

P

Paris Agreement, 21
Partnership, 20–21, 26
Political stability, 21
Polluters, 89
Pollution, 3, 6, 9, 14, 18, 94, 111, 124, 138, 140,
    165
    control, 54
    reduction, 111
    source, 111
Population, 3–4, 46, 48, 77, 79, 82–83, 86, 93
Primary energy, 134
    consumption, 52, 65
        ratio, 144
    index, 155
    intensity, 55
    use, 55
Private sector, 20–21, 138
Public transport, 91–92, 95, 98, 123
    alternatives, 118
    infrastructure, 91, 93
    planning, 115–116, 119
        issues, 114, 117
        studies, 114
    services, 92, 94, 116
    system, 91, 119–120
PV cells, 135

Q

Quality of life, 3, 14, 20, 47, 111, 124, 134
    aspects, 134
    conditions, 136

R

Railways, 118
Reliability, 19–20, 33, 55, 111, 143, 167
Responsibilities, 74
Retailers, 56
Road transport, 90–91, 94, 111, 123, 167
    data, 94
    decarbonizations, 89
    efficiency, 97
    indicators, 91–93
    polluters, 89
    sector, 95
    sustainability, 88, 91, 95–97, 120, 167
        assessment, 88
        issues, 88

S

Sensitivity, 56
    analysis, 74, 81, 83–85, 95, 146, 150–152
SO_2 emissions, 79, 144

Solar, 6, 8, 133, 135–136, 138, 155
  collectors, 135
  electricity, 139
  energy, 79, 135–136, 138–139
    technologies, 136, 138
  heat, 86
  power, 139
  radiation, 15
Solid fuel, 48
Southeast Europe, 48, 80
Stakeholders, 7, 25–29, 31–32, 36, 111–112, 115,
    117, 119–124
Straw, 6
Subsidy, 141
Sustainability indicators, 7
Sustainable development, 1–4, 6–10, 17, 33, 45,
    54, 153, 165
  agenda, 10
  scenarios, 11
  themes, 17
  approach, 32
  goal, 2–3, 9, 11–12, 45, 145–146, 150, 165
  paradigm, 9, 12–14
  concept, 12–14, 167
Sustainable energy, 1–2, 4, 11, 133
  development, 2, 4–21, 35, 48, 51, 54–55, 59–60,
    111, 133, 165, 167
    debate, 15
    goals, 17, 19
    indicators, 166
    issues, 37
    objectives, 55–56
    paradigm, 11–12
    problems, 11
    themes, 17
  dimensions, 8
  production, 10
  consumption, 16–18
    issues, 16
    options, 20
  policy, 19–20, 67
  technologies, 21
  studies, 35
  systems, 45, 67, 137
  sector, 77, 94
  supply, 133
  generation, 139
  options, 167

concept, 4, 9, 11–12
development, 2, 4–21, 35, 48, 51, 54–55, 59–60,
    111, 133, 165, 167
indicator, 7, 13
issues, 10, 30–31
Sustainable future, 7
Sustainable technologies, 73
Synchronization, 64

T

Tax, 3, 21, 68, 82, 87, 90, 93, 97
  incentives, 21, 155
  policy, 93
TOPSIS technique, 39, 70–71, 81, 83, 118, 120,
    138, 146, 156
Transformation, 3, 15, 18, 73
  initiatives, 21
  losses, 68–69
Transition countries, 20
Transparency, 26, 34, 123
Transport sector, 47, 89–90, 94, 111, 115, 119,
    123–124
  assessment, 122
  sustainability, 88, 94–95, 97
    assessment, 120

U

Unsustainable, 2
  energy production, 11

W

Waste, 6, 55, 64, 76, 82–83, 86–88, 137
  management, 54
  water, 6, 79
Wastewater, 80
Well-being, 2–4, 17, 33, 47, 83, 94, 165
  issues, 83
Wind, 6, 8, 135–136, 139, 147–148, 154–155
  energy, 61, 133, 136, 139
  speed, 15
  technology, 136
  turbine, 136
World Energy Council, 4, 11

Z

Zero-carbon energy, 7

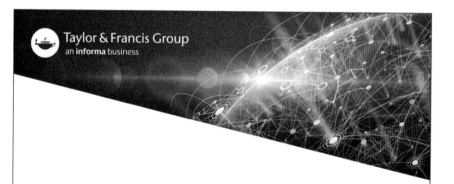